Commendations for Griffith's treatise

From Thought Leaders

'[**Prof. Stephen Hawking**] is most interested in your impressive proposal.'
● 'In all of written history there are only 2 or 3 people who've been able to think on this scale about the human condition.' **Prof. Anthony Barnett**, zoologist ● '*FREEDOM* is the book that saves the world...cometh the hour, cometh the man.' **Prof. Harry Prosen**, former Pres. Canadian Psychiatric Assn. ● 'I am stunned and honored to have lived to see the coming of "Darwin II".' **Prof. Stuart Hurlbert**, esteemed ecologist ● 'Living without this understanding is like living back in the stone age, that's how massive the change it brings is!' **Prof. Karen Riley**, clinical pharmacist ● 'Frankly, I am blown away by the ground-breaking significance of this work.' **Prof. Patricia Glazebrook**, philosopher ● 'I've no doubt a fascinating television series could be made based upon this.' **Sir David Attenborough** ● '*FREEDOM* is the necessary breakthrough in the critical issue of needing to understand ourselves.' **Prof. David J. Chivers**, former Pres. Primate Society of Britain ● 'Whack! Wham! I was converted by Griffith's erudite explanation for our behaviour.' **Macushla O'Loan**, *Executive Women's Report* ● 'This is indeed impressive.' **Dr Roger Lewin**, preeminent science writer ● 'I have recommended Griffith's work for his razor-sharp biological clarifications.' **Prof. Scott Churchill**, psychologist ● 'An original and inspiring understanding of us.' **Prof. Charles Birch**, zoologist ● 'The insights are fascinating and pertinent and must be disseminated.' **Dr George Schaller**, preeminent biologist ● 'Very impressive, particularly liked the primatology section.' **Prof. Stephen Oppenheimer**, geneticist, author *Out of Eden* ● 'I consider the book to be the work of a prophet.' **Dr Ron Strahan**, former dir. Sydney Taronga Zoo ● 'The scholarly value [of Griffith's synthesis] is comparable to several of the most celebrated publications in biology.' **Prof. Walter Hartwig**, anthropologist ● 'I believe you're on to getting answers to much that has bewildered humans.' **Dr Ian Player**, famous Sth. Afr. conservationist ● 'A superb book, a forward view of a world of humans no longer in naked competition.' **Prof. John Morton**, zoologist ● 'This might bring about a paradigm shift in the self-image of humanity.' **Prof. Mihaly Csikszentmihalyi**, psychologist ● 'As a therapist this is a simply brilliant explanation.' **Jayson Firmager**, founder of *Holistic Therapist Magazine* ● 'The questions you raise stagger me into silence; most admirable.' **Ian Frazier**, author *Great Plains* bestseller ● 'The WTM is an island of sanity in a sea of madness.' **Tim Macartney-Snape**, world-leading mountaineer & twice Order of Australia recipient

Commendations From The General Public

'Griffith should be given Nobel prizes for peace, biology, medicine; actually every Nobel prize there is!' ● 'He nailed it, nailed the whole thing, just like the world going from FLAT to ROUND, BOOM the WHOLE WORLD CHANGES, no joke.' ● '*FREEDOM* will be the most influential, world-changing book in history, and time will now be delineated as BG, before Griffith, or AG, after Griffith.' ● 'I'm speechless – this is bigger than natural selection & the theory of relativity!' ● 'I really think this man will become recognized as the best thinker this world's ever seen, and don't we need him right now!' ● 'Griffith has decoded the human species, we FINALLY know what's going on & the suffering stops!' ● 'The world can't deny this for much longer, let the light in, save the human race!' ● 'This is the most exciting moment in my life. *THE Interview* tore my hat off & let my brain fly into the sky!' ● '*THE Interview* should be globally broadcast daily. The healing explanation humans so sorely need.' ● 'In a world that's lost its way there's no greater breakthrough, water to a world dying of thirst.' ● 'Dawn has come at Midnight! A brilliant exposition, we could be on the cusp of regaining Paradise!' ● 'This man has broken the great silence, defeated our denial, got the truth up, woken us from a great trance.' ● 'Beware the 'deaf effect; your mind will initially resist the issue of our corrupted condition and so find it hard to take in or hear what's being said, but if you're patient you'll find the redeeming explanation of our condition pure relief.' ● 'John Lennon pleaded "just give me some truth", well this site finally gives us *all* the truth!' ● '*FREEDOM* is the most profound book since the Bible, now with the redeeming truth about us humans.' ● '*Death by Dogma* is brilliant clarification.' ● 'We were given a computer brain, but no program for it; but Aha, Griffith has found it, made sense of our lives!' ● 'This just goes deeper & deeper in explaining us, like dawn devouring darkness, amazing!' ● 'Agree, this is not another deluded, pseudo idealistic, PC, 'woke', false start to a better world, but the human-condition-resolved real solution.' ● 'Freedom indeed! What we have here is the second coming of innocence who exposes us but sets us free!' ● 'As prophesised, King Arthur has returned to save us (mentioned in par.1036 *Freedom*)' ● 'We all need to go back to school & learn this truthful explanation of life.' ● 'Join in our jubilation, your magic reunites, all men become brothers, all good all bad, be embraced millions! This kiss [of understanding] for the whole world' – From Beethoven's 9th (par.1049 *Freedom*)

THE HUMAN CONDITION

What exactly is it, what caused it, and how the human race has finally liberated itself from the horror of it

Jeremy Griffith

www.HumanCondition.com

The Human Condition: What exactly is it, what caused it, and how the human race has finally liberated itself from the horror of it by Jeremy Griffith

Published in 2025, by WTM Publishing and Communications Pty Ltd (ACN 103 136 778) (www.wtmpublishing.com).

All enquiries to:
WORLD TRANSFORMATION MOVEMENT®
Email: info@worldtransformation.com
Website: www.humancondition.com or www.worldtransformation.com

The World Transformation Movement (WTM) is a global not-for-profit movement represented by WTM charities and centres around the world.

ISBN 978-1-74129-113-1
CIP – Biology, Philosophy, Psychology, Health

FIX THE WORLD™

HumanCondition.com

With the real problem of the human condition finally solved we can now ACTUALLY fix the world!

Contents

Background

Jeremy Griffith is an Australian biologist who has dedicated his life to bringing redeeming and psychologically healing biological understanding to the dilemma of the human condition—which is the underlying issue in all human life of our species' extraordinary capacity for what has been called 'good' and 'evil'.

Jeremy has published over 20 books on the human condition, including:

— *Beyond The Human Condition* (1991), his widely acclaimed second book;

— *A Species In Denial* (2003), an Australasian bestseller;

— *FREEDOM: The End Of The Human Condition* (2016), his definitive treatise;

— *THE Interview* (2020), the transcript of acclaimed British actor and broadcaster Craig Conway's world-changing and world-saving interview with Jeremy about his book *FREEDOM*;

— *Death by Dogma: The biological reason why the Left is leading us to extinction, and the solution* (2021), which presents the biological reason why Critical Theory threatens to destroy the human race;

— *The Great Guilt that causes the Deaf Effect* (2022), which describes how lifting the great burden of guilt from the human race initially causes a 'Deaf Effect' difficulty taking in or 'hearing' what's being presented;

— *The Shock Of Change that understanding the human condition brings* (2022), which addresses how to manage the shock of change that inevitably occurs when the redeeming understanding of our corrupted condition arrives;

— *Therapy For The Human Condition* (2023), which is about the therapy that is desperately needed to rehabilitate the human race from our psychologically upset state or condition, elaborating on what is presented in *FREEDOM*;

— *Our Meaning* (2023), which explains how being able to know and fulfil the great objective and meaning of human existence finally ends human suffering;

— *The Great Transformation: How understanding the human condition actually transforms the human race* (2023), which gives a concise description of how the psychological rehabilitation of humans occurs, and how everyone's life can immediately be transformed;

— *AI, Aliens & Conspiracies: The Truthful Analysis* (2023), which provides Jeremy's thoughts on the much discussed question of the danger of Artificial Intelligence (AI), and on the possibility of alien life visiting Earth, and also his explanation for the epidemic of conspiracy theories; and

— *Sermon On The Beach* (2024), which is Jeremy's elaborated transcript of his inspired description of how the human race now leaves the horror of the human condition forever!

This 2025 book ***The Human Condition: What exactly is it, what caused it, and how the human race has finally liberated itself from the horror of it*** is, as a Director of the World Transformation Movement (WTM), Tony Gowing, has commented, "one of Jeremy's three most important presentations on the human condition, serving as a powerful short-in-length bridge between the other two—his very short introduction in *THE Interview* and his comprehensive presentation in *FREEDOM*. It also takes his human-race-saving insights to a new depth of clarity—so it is a MUST READ!"

Jeremy's work has attracted the support of such eminent scientists as the former President of the Canadian Psychiatric Association Professor Harry Prosen, the esteemed American ecologist Professor Stuart Hurlbert, Australia's Templeton Prize-winning biologist Professor Charles Birch, the former President of the Primate Society of Great Britain Professor David Chivers, world-leading physicist Stephen Hawking, as well as other distinguished thinkers such as the pre-eminent philosopher Sir Laurens van der Post.

Jeremy is the founder and a patron of the WTM, a not-for-profit organisation that makes all its information and material available free of charge—see www.HumanCondition.com.

Jeremy Griffith in 2020

Brief 11-paragraph summary of the fabulous good news about the human condition

[1] Firstly, if you haven't already done so, it is highly recommended that you watch or read *THE Interview* at the top of our World Transformation Movement website (www.humancondition.com), as this 2025 book, *The Human Condition*, is an elaboration on what is introduced in that interview. Indeed, as a Director of the World Transformation Movement, Tony Gowing, has said about the context of this book, it is **'one of Jeremy Griffith's three most important presentations on the human condition, serving as a powerful short-in-length bridge between the other two—his very short introduction in *THE Interview* and his comprehensive presentation in *FREEDOM: The End Of The Human Condition*.'**

———————

[2] The human condition is the underlying problem in all human affairs but it has been such a terrifying subject for us humans to face that the only way we have been able to cope with it has been to deny the problem even exists. Most fabulously, however, what has happened is that through the advances that have been made in science the human condition has finally been able to be solved and the human race liberated from the horror of it. So what exactly then is the human condition, and how is it solved and the whole human race transformed from having to live with the agony and horror of that condition?

[3] In short, the human condition is the psychologically upset angry, egocentric and alienated way we humans have been living, and the resolution of the human condition occurs because we can finally understand and relieve the underlying cause of that psychologically distressed state or condition.

[4] To look more closely at what has just been said. The truth is that despite our species' magnificent mental capabilities and undeniable capacity for immense sensitivity and love, behind every wondrous scientific achievement, sensitive artistic expression and compassionate

act lies the shadow of humanity's darker side—an unspeakable history of greed, hatred, rape, torture, murder and war. And the eternal question has been *what* is the cause of our destructive, insensitive and cruel side? Religious assurances—such as the idea that although we are 'fallen' and supposedly 'sinful' and 'evil', 'God still loves us'—were comforting, but ultimately we needed to know why we became 'fallen' and why we are lovable. *Why* do we thinking, rational, immensely clever, supposedly sensible beings behave so abominably and cause so much suffering and devastation? What is the origin of the dark, volcanic forces that undeniably exists within us humans? What is it deep within us humans that has troubled us so terribly? What is it that makes us such a combative, ruthless, hateful, retaliatory, violent, in truth psychologically disturbed creature?

[5]Most wonderfully relieving for us humans, this question that has been so troubling for us of why are we the way we are, capable of deeds of shocking violence, depravity, indifference and cruelty, has finally been scientifically answered—and in the most psychologically redeeming, reconciling and rehabilitating way!

[6]This **Brief Summary** of what that answer is, which will be explained more fully in Parts 1 and 2 of this book, is actually a very obvious answer once the different natures of the two main influences of our behaviour of our instincts and intellect are considered, and that obvious answer is summarised in the following three sentences. The ability that science has given us to understand that our gene-based, naturally selected instincts were only able to give us orientations to the world means we can now appreciate that when our nerve-based, self-managing, fully conscious mind emerged and began carrying out experiments in managing our life from a basis of understanding the world, our inflexible, 'dictatorial', non-understanding instinctive orientations were inevitably going to be intolerant of and condemning of our conscious mind's instincts-defying, self-managing experiments in understanding, which is our necessary search for knowledge. But without this reconciling and redeeming explanation that science has finally made possible of the difference between the gene and nerve-based learning systems, which is that genes can orientate but nerves need to understand, our conscious mind had no way to counter this feeling of condemnation and criticism from our instincts, other than to attack the 'criticism', try to prove it undeserved, and to block it from its thoughts. This was the genesis of our immensely heroic, psychologically upset, angry, egocentric and alienated HUMAN CONDITION!

[7] What is so indescribably wonderful is that this 'Instinct vs Intellect', 'genes can orientate but nerves need to understand', obvious (now that it has been explained) GOOD reason for our angry, egocentric and alienated, 'good and evil'-stricken, instinctive-self-or-soul-corrupted, 'fallen' human condition has finally lifted the so-called 'burden of guilt' from the human race and ended the need for our defensive angry, egocentric and alienated behaviour, thereby bringing about the psychologically redeemed, reconciled and healed transformation of our species! As Professor Harry Prosen, a former president of the Canadian Psychiatric Association, said of this breakthrough, **'I have no doubt that Australian biologist Jeremy Griffith's instinct vs intellect explanation of the human condition is the holy grail of insight we have sought for the psychological rehabilitation of the human race.'**

[8] The World Transformation Movement (WTM) is the global not-for-profit charity that was established to promote this human-race-saving, **'holy grail'** breakthrough in science of the 'Instinct vs Intellect' resolution of the human condition that ends human conflict and suffering at its source and provides the now urgently needed road map for the complete transformation of our lives and world!

[9] And it should be mentioned that to help everyone access this breakthrough, the WTM doesn't charge for subscription or membership, and provides all books, videos, essays and other materials about the human condition free of charge on our website.

---

[10] To reiterate, when our conscious mind emerged, its necessary experiments in managing our life from a basis of understanding were condemned by our instinctive orientations to the world because, in undertaking those experiments, our conscious mind was not behaving in accordance with those orientations, and then that condemnation upset our conscious mind, causing it to become defensively angry towards the criticism, determined to prove it undeserved, and committed to blocking it out. We—our conscious thinking mind—became psychologically upset angry, egocentric and alienated sufferers of the human condition. But what is indescribably wonderful is that through having the redeeming understanding of why this condition emerged, the need for our defensive angry, egocentric and alienated behaviour it produced ends, thus bringing about the psychological rehabilitation of the human race. We are transformed

from having to suffer from the agony and horror of the human condition to living free of it in psychologically relieved peace, togetherness and happiness at last!

[11] So, the human condition is the immensely mentally insecure and psychologically distressed state that all humans have suffered from as a result of a clash between our species' original non-understanding, dictatorial, instinctive *orientations* to the world, and our newer fully conscious mind that needs to *understand* the world to operate. And the World Transformation Movement is the disseminator of this psychologically relieving and transforming, Instinct vs Intellect, **'holy grail of insight'** explanation of the human condition—the insight humanity has tirelessly worked toward, yearned for, and now so desperately needs!

Part 1

What exactly is the human condition, what caused it, and how the human race has finally liberated itself from the horror of it

Part 1.1 Our great fear of the subject of the human condition has led to evasive interpretations of what it is

[12] What this Part 1 presents is a full explanation of what exactly the 'human condition' is, why we humans have been living in mortal fear of it, and of the scientific advances that have finally enabled us to free ourselves from the agony and horror of the human condition.

[13] Firstly, to look at different interpretations of the human condition.

[14] As will be revealed, the human condition has been an unbearably depressing subject for us humans to face while we couldn't truthfully explain it. In fact, the human condition has been such a deep, serious, and guilt-ridden issue that we have been scared to think about it; I have even heard 'the human condition' referred to as **'the personal unspeakable'** and as **'the black box inside of us humans that we can't go near'**. So the human condition is a *very* deep, serious and foreboding subject. However, as we will see, through finally having what the human condition really is properly and compassionately explained and understood, it is, thank goodness, no longer a terrifying, **'can't go near it'** subject—and, best of all, the problem it refers to is totally solved by that explanation, which means the human race can become completely peaceful, loving, calm and happy! If you don't believe me, please read on!

[15] The following interpretations of the human condition vary in their degree of evasion of the once—but no longer—unbearably depressing and confronting subject of the human condition.

[16] The most evasive interpretation of the 'human condition' is to say it is about the physical and practical hardships and struggles to survive in human life, like having to find a suitable place and dwelling to live in, or having at times to endure natural disasters like floods and droughts, or having to cope with famine or poverty or remoteness or isolation.

However, such physical and practical hardships of human existence are not at all what the human condition really is. The human condition is not a physical struggle but a mental struggle. It refers to a question we wrestle with in our mind.

[17]Less evasive interpretations of the human condition do recognise that it has something to do with a mental struggle, but these guilt-free accounts still find a way to avoid going anywhere near what that struggle really is. Such interpretations include saying that the human condition is the despair that comes with being conscious because it makes us realise we will die someday; or that the human condition is simply the terror we conscious humans experience from having the free will to choose what to do without knowing the outcome.

[18]A much more accurate—but in fact, as we will see, still extremely inadequate—interpretation of what the term 'human condition' means is that it refers to the anguish caused by our inability to answer the question of why, if the universally accepted ideals of life are to be cooperative, selfless and loving—ideals that have been accepted by modern civilisations as the foundations for constitutions and laws and by the founders of all the great religions as the basis of their teachings—have we humans been competitively, selfishly and aggressively behaved?

[19]However, this interpretation also doesn't, in truth, go anywhere near what we humans are talking about when, in moments of profound and very serious thought, we dared to refer to the 'human condition'. After all, we have what we consider an obvious explanation for our competitive, selfish and aggressive behaviour—that it is due to us having must-reproduce-our-genes, 'survival of the fittest', savage instincts like other animals. We think, *'Other animals are always fighting and aggressively competing with each other to make sure they pass their genes on—as the process of natural selection dictates they must try to do—and of course that is our biological heritage as well. We have competitive, selfish and aggressive instincts in us that emanate from our animal past that we are always having to try to contain and restrain. That's the basic responsibility and task of being a rational, sensible, conscious thinking human—to make sure our competitive, selfish and aggressive animal instincts don't get out of control!'* So having that seemingly obvious excuse for our competitive, selfish and aggressive behaviour indicates that when in moments of profound thought we dared to refer to the 'human condition' we must have been referring to something far deeper and much more serious, which, as we will see, it certainly is.

Part 1.2 The true interpretation of the human condition

[20] Identifying what the human condition *really* is begins by looking more closely at this must-reproduce-our-genes, 'survival of the fittest', 'savage instincts' explanation that we have been using for our competitive, selfish and aggressive behaviour.

[21] It is certainly true that we have been comfortably attributing our competitive, selfish and aggressive behaviour to us humans having must-reproduce-our-genes, 'survival of the fittest', savage instincts like other animals. This 'savage instincts' reason is what we have been taught at school and what we hear in every documentary. Indeed, everywhere in articles we read and in our conversations we see and hear comments like: 'We are programmed by our genes to try to dominate others and be a winner in the battle of life'; and 'Our preoccupation with sexual conquest is due to our primal instinct to sow our seeds'; and 'Men behave abominably because their bodies are flooded with must-reproduce-their-genes-promoting testosterone'; and 'We want a big house because we are innately territorial'; and 'Fighting and war is just our deeply-rooted combative animal nature expressing itself'. (I should mention that, as I explain in Part 9 of my book *Death by Dogma*, and in other presentations, while left-wing thinkers use the biologically impossible 'group selection' theory to claim we have some cooperative, selfless and loving instincts, they also say we have this competitive, selfish and aggressive 'animal' side, which the political philosopher Karl Marx limited to such basic needs as sex, food, shelter and clothing. To quickly explain 'group selection', while it is a seemingly plausible theory that a group that is cooperative will always be more successful and so reproduce their genes more than a group whose members compete with each other, this idea is actually false biology because, as the biologist Jerry Coyne pointed out, of **'the tendency of each group to quickly lose its altruism through natural selection favoring cheaters** [selfish, opportunistic individuals]' ('Can Darwinism improve Binghamton?', *The New York Times*, 9 Sep. 2011).) Yes, 'you can help me reproduce my genes but I'm not about to help you reproduce yours' is the biological reality of the natural selection process.

byrdyak - stock.adobe.com;
jeaneeem / Flickr

[22] Most significantly however, as I point out in Part 1 of *THE Interview* I did with the British broadcaster and actor Craig Conway (which appears at the top of our World Transformation Movement homepage—and I will be suggesting various documents like this that provide helpful elaborations on what is being described as I progress through this book), this must-reproduce-our-genes, 'savage instincts' excuse cannot be the real reason for our species' competitive, selfish and aggressive behaviour because, after all, words used to describe human behaviour such as egocentric, arrogant, inspired, depressed, deluded, pessimistic, optimistic, artificial, hateful, cynical, mean, sadistic, immoral, brilliant, guilt-ridden, evil, psychotic, neurotic and alienated, all recognise the involvement of *our* species' fully conscious thinking mind. They demonstrate that there is a *psychological* dimension to our behaviour; that we don't suffer from a genetic-opportunism-driven 'ANIMAL CONDITION', but a conscious-mind-based, *psychologically* troubled HUMAN CONDITION—the source of which is what is going to be explained.

[23] To clarify, according to the *Cambridge Dictionary*, 'psychological' means **'relating to the human mind and feelings'**. So yes, we suffer from a **'mind and feelings'** situation, which, as we progress with this description of what the human condition *really* is, will be explained as our self-managing, knowledge-finding, fully conscious **'mind'** being upset by criticising **'feelings'** from our non-understanding, dictatorial instincts. But, as I say, that is yet to be explained—but yes, we suffer from a fully-conscious-**'mind'**-involved-with-**'feelings'** *psychological* human condition, not a relatively-unconscious-mind-with-instincts-in-control animal condition.

[24] Further, this idea that we have savage, must-reproduce-our-genes, competitive, selfish and aggressive instincts like other animals can't be true because we humans actually have cooperative, selfless and loving *moral* instincts, the 'voice' or expression of which we call our conscience, which is the complete *opposite* of competitive, selfish and aggressive instincts! As Charles Darwin said, **'The moral sense...affords the best and highest distinction between man and the lower animals'** (*The Descent of Man*, 1871, ch.4). Of course, to have cooperative, selfless and loving moral instincts our distant ape ancestors must have *lived* cooperatively, selflessly and lovingly, otherwise how else could we have acquired them? Our ape ancestors can't have been brutal, club-wielding, competitive, selfish and aggressive savages as we have been taught, rather they must have lived in a Garden of Eden-like state of cooperative, selfless and loving innocent gentleness—which, as will shortly be evidenced, is a state the bonobo

species of ape is currently living in, and which anthropological findings now evidence our species *did* once live in. For instance, anthropologists like C. Owen Lovejoy have reported that **'our species-defining cooperative mutualism can now be seen to extend well beyond the deepest Pliocene** [which is well beyond 5.3 million years ago]' (*Science*, 2009, Vol.326, No.5949).

Group of bonobos

'Breakthrough of the Year': cover of the December 2009 issue of *Science* magazine

Matternes's reconstruction of the 4.4 mya *Ardipithecus ramidus* in its natural habitat

[25] So, saying our competitive, selfish and aggressive behaviour comes from humans having savage, must-reproduce-our-genes, competitive,

selfish and aggressive instincts is simply not true. What will become clear is that this was just a convenient excuse we used while we waited for the real explanation for our conscious-mind-based, deeply *psychologically* troubled human condition—which is what is going to be presented. I should also mention that shortly in Part 1.4 it will be explained how it was the process of prolonged nurturing that enabled our ape ancestors to overcome the genetic imperative to reproduce our genes and by so doing become cooperatively, selflessly and lovingly behaved. This nurturing is also how the bonobos were able to become so harmoniously behaved, as the nurturing mother bonobos in the photo above provides some evidence of. Again, this will all be explained in Part 1.4.

[26] Coming back then to the question of what the human condition *really* refers to: being competitive, selfish and aggressive is what other animals are; what we humans are is far worse than just being competitive, selfish and aggressive because what *we* are is something seriously sinister! As has been pointed out, despite our species' magnificent mental capabilities, and undeniable capacity for immense sensitivity and love, behind every wondrous scientific achievement, sensitive artistic expression and compassionate act lies the shadow of humanity's darker side—an unspeakable history of greed, hatred, rape, torture, murder and war; a propensity for deeds of shocking violence, depravity, indifference and cruelty. It is true—undermining all our marvellous accomplishments and sensibilities is the fact that we humans have also been the most ferocious and malicious creatures to have ever lived on Earth! As the philosopher Arthur Schopenhauer wrote, '**man is the only animal which causes pain to others with no other object than causing pain…No animal ever torments another for the sake of tormenting: but man does so, and it is this which constitutes the *diabolical* nature which is far worse than the merely bestial**' (*Essays and Aphorisms*, tr. R.J. Hollingdale, 1970, p.139 of 237).

[27] Yes, being competitive, selfish and aggressive, which is what other animals are, doesn't come anywhere near to the heart of the issue about us humans of why we are the way we are, which is *psychologically* distressed; in fact, *psychologically* deranged. And the eternal question has been *why*? What is the reason for our dual capacity for what is referred to in Genesis in the Bible as 'good and evil' in our make-up? Are we essentially 'good' and, if so, what is the cause of our dark, destructive, insensitive, cruel, seemingly 'evil' side? As mentioned, religious assurances—such as the idea that although we are 'fallen' and supposedly 'sinful' and 'evil', 'God still loves us'—offered comfort, but ultimately we needed to know

why we became 'fallen' and why we are lovable. *Why* do we thinking, reasoning, rational, immensely clever, supposedly sensible beings behave so abominably and cause so much suffering and devastation? <u>What is the origin of the dark, volcanic forces that undeniably exist within us humans? What is it deep within us humans that has troubled us so terribly? What is it that makes us such combative, ruthless, hateful, retaliatory, violent, in truth psychologically disturbed beings?</u>

[28] Significantly, following a request in 1988 from *TIME* magazine to the renowned South African author Alan Paton to submit an essay on apartheid in South Africa, *TIME* instead received from Paton, and published in its place, a deeply reflective article on Paton's favourite pieces of literature. In what proved to be the writer's last publication, Paton wrote: **'I would like to have written one of the greatest poems in the English language—William Blake's "Tiger, Tiger Burning Bright", with that verse that asks in the simplest words the question which has troubled the mind of man—both believing and non believing man—for centuries: "When the stars threw down their spears / And watered heaven with their tears / Did he smile his work to see? / Did he who made the lamb make thee?"'** (25 Apr. 1988).

[29] The reason Blake's 1794 poem *The Tyger* resonated so strongly with Paton, and is one of the most famous poems in the English language (it has been described as **'the most anthologized poem in English'** (*The Cambridge Companion to William Blake*, ed. Morris Eaves, 2003) and is a mainstay of the English curriculum in schools) is because of its profundity. The poem raises that fundamental question involved in being human of how could the nasty, mean, cruel, indifferent, 'dark side' of our psychological makeup—represented by the **'Tiger'**—be both reconcilable with and derivative of the same force that created **'the lamb'** in all its innocence? (More from Blake's poem *The Tyger* will be included and explained in paragraphs 130-131.)

Maria Lindsey/Pexels

[30]Yes, that is the *real* question about us humans: *why* have we been not just competitive, selfish and aggressive, but *so* ferociously aggressive and brutal, *so* extremely self-preoccupied and selfish, and *so* deeply unaware and estranged from being able to feel for others—in fact, *so* *psychologically* angry, egocentric and alienated that life has become all but unbearable and we have nearly destroyed our own planet? Are we essentially bad, a flawed species, an evolutionary mistake, a blight on Earth, a cancer in the universe? Or could we possibly be sublimely wonderful (which is actually what we are going to be revealed as being). And, more to the point, is the human race faced with the prospect of having to live forever in this tormented state of uncertainty and insecurity about the fundamental goodness, worth and meaning of our lives? Is it our species' destiny to have to live in a state of permanent damnation?! (As we will also see, most fabulously we *ARE* going to be able to free ourselves from this dreadful state!)

[31]While religious assurances such as 'God loves you' may have provided temporary comfort, they failed to explain *WHY* we are lovable. So, *WHY* are we lovable? How could we humans be good when all the evidence seems to unequivocally indicate that we are a deeply flawed, bad, even evil species? What is the answer to this problem of 'good and evil' in the human make-up? What caused us to become so brutally divisively behaved? The agony of having been unable to truthfully (i.e., not resort to the 'savage instinct' excuse) explain our obviously extremely psychologically distressed lives has been a particular affliction and burden of human life, our seemingly horrifically flawed state or condition—our psychologically distressed 'HUMAN CONDITION'.

[32]So this is what 'the human condition' really is: our *horrifically* psychologically distressed, angry, egocentric and alienated state or condition—the biological origins and resolution of which will be presented. However, in order to do so, it is first necessary to explain two particular features of our lives. They are HOW HUMANS ACQUIRED OUR COOPERATIVE, SELFLESS AND LOVING MORAL INSTINCTS, and HOW HUMANS BECAME A FULLY CONSCIOUS SPECIES WHEN OTHER SPECIES HAVEN'T.

Part 1.3 Clarification of what instincts and consciousness are

[33] In order to explain how we humans acquired our cooperative, selfless and loving moral instincts, and how we became fully conscious when other species haven't, it first needs to be explained what instincts and consciousness actually are.

[34] Firstly, to explain instincts. While animals largely depend on their nervous system to coordinate their movement and control how they react to their environment, other systems such as their hormonal, circulatory, digestive, immune and reproductive systems also influence how they behave. Obviously all these systems that affect how a species of animal moves and behaves have been honed over many generations by natural selection in the course of adapting that species to its environment. These naturally selected genetic traits that orientate an animal's movements and behaviour are referred to as its instincts. Animals move about and behave in many different ways—they fight and 'court' each other, they build nests, they search for food, they migrate, etc—and natural selection has given them genetic programming, 'instincts', to control and orientate all this movement and behaviour.

[35] In the case of consciousness, as is described in detail in chapter 7 of *FREEDOM*, and summarised in Freedom Essay 24 (both of these documents, and indeed all my presentations, are freely available at our World Transformation Movement website at www.humancondition.com), nerves were, as just mentioned, originally developed to coordinate movement in animals. Significantly, there is one aspect of nerves' ability to control how animals react to their environment that has the potential to give rise to consciousness—and this is an aspect that is largely independent of any instinctive orientations of an animal's nervous system that developed through natural selection. This aspect of the nervous system that gave rise to the potential to develop a conscious understanding of cause and effect is nerves' ability to store impressions—what we refer to as 'memory'. An electric current passed through a nerve leaves an imprint of its passage, which represents a memory of that piece of information. This ability to remember past events makes it possible to compare them with current events and identify regularly occurring experiences. This knowledge of, or insight into, what has commonly occurred in the past makes it possible

to predict what is likely to happen in the future and to adjust behaviour accordingly. Once insights into the nature of change are put into effect, the self-modified behaviour starts to provide feedback, refining the insights further. Predictions are compared with outcomes and so on. Much developed—and such refinement occurred in the human brain—nerves can sufficiently *associate* information to *reason* how experiences are related, learn to *understand* and become CONSCIOUS of, or aware of, or *intelligent* about, the relationship between events that occur through time. Thus consciousness means being sufficiently aware of how experiences are related to attempt to manage change from a basis of understanding.

[36] It should be pointed out that admitting the above obvious explanations for what instincts actually are and what consciousness actually is has been avoided by human-condition-avoiding Reductionist, Mechanistic science. As will become clear shortly when humans' need to avoid the unbearably depressing issue of the human condition is explained, scientists, being extreme sufferers of the human condition like virtually every other human, have also had to live in determined denial of that unbearably depressing subject of the human condition. To do this they avoided the overarching big issue of the human condition and reduced their focus to only looking down at the mechanisms of the workings of our world. They have been 'Reductionist' and 'Mechanistic', not whole-view-confronting or 'Holistic'. (I should mention that while terms like 'Reductionist', 'Mechanistic' and 'Holistic' will be more properly explained and be more easily understood as this presentation progresses, chapter 2:4 of *FREEDOM*, and my book *Don't Stand In the Way, For the Times Are A-Changin'*, present a complete description of science's practice of avoiding the issue of the human condition, and of the terms Reductionism, Mechanism and Holism.) So—and again this will all be made understandable shortly when humans' need to avoid the unbearably depressing issue of the human condition is explained—Reductionist, Mechanistic scientists have had to avoid what instincts actually are and what consciousness actually is because without the redeeming explanation for why we, our conscious thinking self, corrupted our cooperative, selfless and loving **'moral'** instincts that Darwin said **'affords the best and highest distinction between man and the lower animals'**, any admission that the human race once lived in a cooperative, selfless and loving moral instinctive state, and any admission that our conscious mind corrupted that pure, innocent state, was unbearable. And so to avoid any thought

journey getting underway that might lead to those condemning realisa-
tions, the human-condition-avoiding, 'keepers of the lie', Reductionist,
Mechanistic scientists realised it was best to stop that thought journey
at the outset by claiming we just don't know what instincts actually
are and consciousness actually is. For example, when I tried to explain
the 'Instinct vs Intellect' explanation of the human condition (which is
shortly going to be presented) to the former Chief Scientific Adviser to
the UK government, Lord Robert May (note, 'Lord' is a higher ranking
than 'Sir'), at Oxford University in 2014, he said, **'But Jeremy, we don't
know what instincts actually are, or how we actually got them'** (WTM records, 13
Nov. 2014). Certainly instincts are somewhat complicated, but, as the brief
explanation and description of instincts that I have just given evidences,
they are not nearly as bewildering as Lord May tried to make out. In
the case of consciousness, chapter 7 of *FREEDOM* presents a detailed
description of how consciousness has been deliberately left cloaked in
mystery and confusion—again, because a clear and straight-forward and
obvious description of it, as was just given, brought us too close to the
unbearable truth that our conscious mind had committed the seeming
terrible crime of corrupting our original cooperative, selfless and loving,
innocent instinctive life—which all still has to be explained in this book.
I should acknowledge here for the reader that unfortunately there are so
many new ideas that finally can and have to be introduced now that the
human condition is explained that it is impossible to introduce them all
sequentially.

[37] So while the 'gene-based learning system' does involve the genetic
selection of nerve pathways and networks, what is meant by the 'nerve-
based learning system' in the coming descriptions is the dimension or
aspect of the nervous system that is to do with understanding cause and
effect. This honest but not very explanatory dictionary definition of
instincts makes this particular difference clear—instincts are **'a largely
inheritable and unalterable tendency of an organism to make a complex and
specific response to environmental stimuli without involving reason'** (*Merriam-
Webster Dictionary*; see www.wtmsources.com/144).

[38] As is going to be explained shortly in Part 1.6, the immense significance of the nerve-based learning system becoming sufficiently developed in humans for us to become fully conscious and able to effectively
manage events, was that our conscious intellect was then in a position to

wrest control of our lives from our gene-based learning system's instincts, which, up until then, had been in charge of our lives—the effects of which, as we will see, were catastrophic!

Part 1.4 How we humans acquired our cooperative, selfless and loving moral instincts

[39] So the first of the two instinct and intellect features of our lives that needs to be explained for the human condition to be fully understood and ended is how we humans acquired our cooperative, selfless and loving moral instincts. (The second—how we became fully conscious—will be explained in the next Part.)

[40] A fundamental truth in biology is that genes normally can't select for unconditionally selfless, fully cooperative instinctive traits, simply because such traits tend to be self-eliminating and so normally can't become established in a species—'By all means, you can be selfless and sacrifice your genes for me but I'm not about to be selfless and sacrifice my genes for you' has been the reality of the process of natural selection. Yes, as the biologist Jerry Coyne pointed out (see paragraph 21), the idea that selfless traits can become established in a group is false biology because of **'the tendency of each group to quickly lose its altruism through natural selection favoring cheaters** [selfish, opportunistic individuals]'. So the question is: how could such a selfish process as natural selection have created loving selfless instincts in us?

[41] As I describe in Part 3 of *THE Interview*, and more fully in chapter 5 of *FREEDOM* (and also summarise in Freedom Essay 21), we acquired our moral instincts through nurturing. To appreciate what is so significant about a mother's nurturing of her offspring, it first needs to be explained that a mother's maternal instinct to care for her offspring *is* selfish because she is ensuring the reproduction of her genes by ensuring the survival of offspring who carry her genes. So maternalism is a selfish trait, which, as I've said, genetic traits normally have to be for them to reproduce and carry on into the next generation. HOWEVER, and this is all-important, from the infant's perspective maternalism does have the appearance of being selfless. From the infant's perspective, it is being treated unconditionally selflessly—the mother is giving her offspring food, warmth, shelter, support and protection for apparently nothing in return. So it follows that if

the infant can remain in infancy for an extended period and be treated with a lot of seemingly altruistic love, it will be indoctrinated with that selfless love and grow up to behave accordingly. <u>So selfish maternalism *can* train an infant in altruistic selflessness</u>.

[42] And primates, being semi-upright from living in trees (arboreal), swinging from branch to branch, and thus having their arms free to hold a dependent infant, are especially facilitated to support and prolong the mother-infant relationship, and so develop this nurtured, loving, cooperative behaviour. And in fact, bonobos, the ape species who live south of the Congo River in Africa, are extraordinarily matriarchal, or female role focused, and exceptionally nurturing of their offspring, and remarkably cooperative, selfless and loving—as the following pictures of bonobos and quotes about their behaviour evidence.

Bonobo mothers holding their infants

Frans Lanting/Mint Images/Getty Images

Bonobos nurturing their infants

[43] Bonobo zoo keeper Barbara Bell wrote that **'Adult bonobos demonstrate tremendous compassion for each other...For example, Kitty, the eldest female, is completely blind and hard of hearing. Sometimes she gets lost and confused. They'll just pick her up and take her to where she needs to go...They're extremely intelligent...They understand a couple of hundred words...It's like being with 9 two and a half year olds all day...They also love to tease me a lot...Like during training, if I were to ask for their left foot, they'll give me their right, and laugh and laugh and laugh'** ('The Bonobo: "Newest" apes are teaching us about ourselves', *Chicago Tribune*, 11 Jun. 1998).

[44] Researchers have also reported that **'bonobos historically have existed in a stable environment rich in sources of food...and unlike chimpanzees have developed a more cohesive social structure'** (Takayoshi Kano & Mbangi Mulavwa, 'Feeding ecology of the pygmy chimpanzees (*Pan paniscus*)'; *The Pygmy Chimpanzee*, ed. Randall Susman, 1984, p.271 of 435). For example, **'up to 100 bonobos at a time from several groups spend their night together. That would not be possible with chimpanzees because there would be brutal fighting between rival groups'** (Paul Raffaele, 'Bonobos: The apes who make love, not war', Last Tribes on Earth.com, 2003; see www.wtmsources.com/143).

[45] Primatologist Sue Savage-Rumbaugh said, **'Bonobo life is centered around the offspring. Unlike what happens among chimpanzees, all members of the bonobo social group help with infant care and share food with infants. If you are a bonobo infant, you can do no wrong...Bonobo females and their infants form the core of the group'** (Sue Savage-Rumbaugh & Roger Lewin, *Kanzi: The Ape at the Brink of the Human Mind*, 1994, p.108 of 299).

[46] A filmmaker of the French documentary *Bonobos* observed that **'They're surely the most fascinating animals on the planet. They're the closest animals to man** [in that they share almost 99 percent of our genetic make-up]. **They're the only animals capable of creating the same "gaze" as a human…Once I got hit on the head with a branch that had a bonobo on it. I sat down and the bonobo noticed I was in a difficult situation and came and took me by the hand and moved my hair back, like they do. So they live on compassion, and that's really interesting to experience'** (short accompanying film to the 2011 French documentary *Bonobos*).

[47] And bonobo researcher Vanessa Woods gave this first-hand account of bonobos' unlimited capacity for love from her study of them in their home in the Congo basin: **'Bonobo love is like a laser beam. They stop. They stare at you as though they have been waiting their whole lives for you to walk into their jungle. And then they love you with such helpless abandon that you love them back. You have to love them back'** ('A moment that changed me – my husband fell in love with a bonobo', *The Guardian*, 1 Oct. 2015).

[48] I should point out, as I did in Part 3 of *THE Interview*, that <u>we couldn't admit how nurturing was able to create cooperative, selfless and loving behaviour while we couldn't explain how our conscious thinking mind became so upset and angry, egocentric and alienated, and as a result, so competitively, selfishly and aggressively behaved (which is what is all about to be explained), and as a result of that development lost the ability to adequately nurture our offspring with unconditional love</u>. As I said in *THE Interview* about this situation, citing the lauded Australian writer and school teacher John Marsden, **'parents would rather admit to being an axe murderer than a bad mother or father'**! (*Sunday Life*, *The Sun-Herald*, 7 Jul. 2002). So, how <u>nurturing gave us our cooperative, selfless and loving moral instincts</u> is another of the critically important truths we couldn't admit while we couldn't present the explanation that is about to be given for our psychologically distressed, competitive, selfish and aggressive human condition—which, as I said in Part 3 of *THE Interview*, is why philosopher John Fiske's 1874 recognition of how nurturing created our moral instincts has been ignored by human-condition-avoiding Reductionist Mechanistic science.

[49] <u>With regard to our own species' time living in a cooperative, selfless and loving innocent state</u>, as I also mentioned in Part 3 of *THE Interview*, way back in about 800 BC, the Greek poet Hesiod acknowledged this time

in his epic poem *Works and Days* (the underlinings in quotes are my emphasis): 'When gods alike and mortals rose to birth / <u>A golden race</u> the immortals formed on earth…Like gods they lived, with <u>calm untroubled mind</u> / Free from the toils and anguish of our kind / Nor e'er decrepit age misshaped their frame…Strangers to ill, their lives in feasts flowed by…Dying they sank in sleep, nor seemed to die / Theirs was each good; the life-sustaining soil / Yielded its copious fruits, unbribed by toil / They with abundant goods '<u>midst quiet lands</u> / <u>All willing shared</u> the gathering of their hands' (*The Remains of Hesiod the Ascræan*, tr. C.A. Elton, pp.17-18).

[50] As I also included in Part 3 of *THE Interview*, Hesiod's Greek compatriot Plato gave this similar description of our species' time in innocence. In 360 BC, which is also a very long time ago, Plato wrote that 'there was a time when…we beheld the beatific vision and were initiated into a mystery which may be truly called most blessed, celebrated by us in <u>our state of innocence, before we had any experience of evils to come, when we were admitted to the sight of apparitions innocent and simple and calm and happy,</u> which we beheld shining in pure light, pure ourselves and not yet enshrined in that living tomb which we carry about, now' (*Phaedrus*; tr. B. Jowett, 1871, 250). I also included Plato's other description of the innocent '**'Golden' Age**' in our species' past, where he wrote of a time when we lived a 'blessed and spontaneous life…[where] neither was there any violence, or devouring of one another, or war or quarrel among them…In those days God himself was their shepherd, and ruled over them [in other words, our original instinctive self was orientated to living in an ideal, godly, cooperative, selfless and loving way]… Under him there were no forms of government or separate possession of women and children; for all men rose again from the earth, having <u>no memory of the past</u> [in other words, we lived in a pre-conscious state]. And…the earth gave them fruits in abundance, which grew on trees and shrubs unbidden, and were not planted by the hand of man. And they dwelt naked, and mostly in the open air, for the temperature of their seasons was mild; and they had no beds, but lay on soft couches of grass, which grew plentifully out of the earth' (*The Statesman*, c.350 BC; tr. B. Jowett, 1871, 271-272).

[51] These references from earlier times, when there wasn't as much psychological upset in the world and thinkers were sounder and able to be more honest, do truthfully acknowledge the cooperative, selfless and loving innocence of our distant ancestors. They also recognised that these ancestors had a 'calm and untroubled mind' and 'no memory of the past'; in other words, they weren't yet conscious. <u>As we are about to see, this development of a fully conscious mind in our distant ancestors (which,</u>

incidentally, the quotes about the bonobos evidence they are well on the way to developing) was immensely significant.

[52] The following sequence of pictures and text summarise this truth that our ape ancestors, like bonobos, were cooperative, selfless and loving nurturers, not savage, barbaric brutes.

Our ape ancestors were innocent, loving, nurturers,

Paleoartist Jay H. Matternes's unusually honest reconstruction of our ancestor, the 4.4 mya *Ardipithecus ramidus*, which appeared in the Dec. 2009 edition of *Science*, which the earlier mentioned report from C. Owen Lovejoy about our past **'co-operative mutualism'** also appeared in. See Freedom Essay 22 for more on the fossil evidence of our nurtured past.

NOT savage, barbaric brutes as they have for so long been portrayed.

It is us humans now who are psychotic angry, egocentric and alienated, psychologically distressed, seemingly 'evil' monsters!

Detail from Jean-Michel Basquiat's 1982 'Untitled' painting which was sold in May 2017 for US$110.5 million, which, at the time was the sixth most expensive artwork ever sold at auction, no doubt because of its extraordinarily honest portrayal of the true nature of our present horrifically psychologically distressed human condition.

[53] So nurturing is how we acquired our cooperative, selfless and loving instinctive self or soul, and, as I have said, these two truths of the importance of nurturing and of the existence of our cooperative, selfless and loving soul couldn't be admitted while we couldn't explain why we corrupted our soul—which is the explanation that is being given in this book.

[54] Interestingly, the truth of the importance of nurturing in creating a sound, secure and well-adjusted human is recognised in Christianity's description of Christ's mother as being the 'virgin mother' (see Matt. 1:23 & Luke 1:26-34)—and it is also recognised in all the images in Christianity of the nurturing, loving 'virgin mother and child', like the drawing I have done opposite. To produce the extraordinarily innocent and sound Christ required an exceptionally innocent and sound nurturing mother, a metaphorically 'un-fucked' virgin mother. The explanation for why sex as humans have been practicing it under the duress of the human condition is actually an attack on women for their lack of empathy for men's immensely heroic but immensely upsetting role in defying our instincts' unjust condemnation in our species' all-important search for knowledge (which is shortly going to be explained) is presented in chapter 8:11B of *FREEDOM* where the different roles of men and women in humanity's journey is described. This underlying truth about sex was recognised by writer and activist Andrea Dworkin when she said, 'All sex is abuse' (*Intercourse*, 1987; reported in *The Sydney Morning Herald*, 25 Jul. 1987; see www. wtmsources.com/172). Importantly however, while sex *has been* an attack on the naivety of women, an act of aggression, it has *also* been one of the greatest distractions and releases of frustration and, on a nobler level, an inspirational act of love, an act of real affection derived from a shared faith in the ultimate meaning of the lives of men and women—which, as is going to be explained, has been to find the redeeming understanding of our corrupted human condition.

Drawing by Jeremy Griffith © 2006 Fedmex Pty Ltd

Jeremy's 2006 tender drawing of the archetypal image of the Madonna and child. Jeremy has said about the image of the Madonna and child: "That this image is such a feature of Christian mythology is powerful recognition that it was the relatively alienation-free unconditional love Christ received from his mother that enabled him to be the exceptional denial-free, truthful thinking prophet that he was. When the great psychiatrist R.D. Laing wrote that **'Each child is a new beginning, a potential prophet'** (*The Politics of Experience* and *The Bird of Paradise*, 1967) he was recognising how much the upset, alienated world of humans today has corrupted our all-loving and all-sensitive instinctive self or soul. As the great playwright Samuel Beckett said about the brevity of innocence in the lives of most humans now: **'They give birth astride of a grave, the light gleams an instant, then it's night once more'** (*Waiting for Godot*, 1955)!"

Part 1.5 The integrative meaning of existence, and how we humans became fully conscious when other species haven't

[55] Having now explained how we humans acquired our cooperative, selfless and loving moral instincts, the second of the two instinct and intellect features of our lives that needs to be explained for the human condition to be understood and resolved is how we humans became fully conscious when other species haven't.

[56] In chapter 7 of *FREEDOM* (and in shorter form in Freedom Essay 24), I explain that the emergence of our fully conscious mind was made possible by us having become cooperatively, selflessly and lovingly behaved (as a result of nurturing), and thus able to think in an Integrative-Meaning-compliant, truthful, effective, meaningful, 'conscious' way.

[57] In order to elaborate on this explanation for the emergence of our fully conscious mind, I need to explain what 'Integrative Meaning' in the term 'Integrative-Meaning-compliant' means. As is explained in chapter 4 of *FREEDOM*, and summarised in Freedom Essay 23, while we couldn't explain our divisive, seemingly *disintegrative* competitive, selfish and aggressive behaviour, the truth of the integrative nature or theme or meaning of our world and of our existence had to be denied. But like the truth about how nurturing created our cooperative, selfless and loving moral instinctive self, this too can now be safely admitted under the umbrella of the about-to-be-presented redeeming explanation for the human condition.

[58] Everywhere we look we see hierarchies of ordered matter: 'There is a tree that is composed of parts (leaves, branches, a trunk and roots) and in turn those parts are composed of parts (fibres, cells, etc).' Our world is clearly composed of a hierarchy of ordered matter: atoms have come together or integrated to form compounds, which in turn have come together or integrated to form virus-like organisms, which in turn came together or integrated to form single-celled organisms that then integrated to form multicellular organisms, which in turn integrated to form societies of single species that continue to integrate to form stable, ordered arrangements of different species. Clearly, what is happening on Earth is that matter is integrating into larger and more stable wholes. And this development of order is not only occurring here, it is also happening out in the universe where, over the eons, a chaotic cosmos continues to organise itself into stars, planets and galaxies. As two of the world's greatest physicists, Stephen Hawking and Albert Einstein, have said, respectively, **'The overwhelming impression is of order…[in] the universe'** (Gregory Benford, 'The time of his life', *The Sydney Morning Herald*, 27 Apr. 2002; see www.wtmsources.com/170), and **'behind everything is an order'** (*Einstein Revealed*, PBS, 1997).

[59] The law of physics that accounts for this integration of matter is known as the 'Second Path of the Second Law of Thermodynamics', or 'Negative Entropy', which states that in an open system, where energy can come into the system from outside it (in Earth's case, from the sun, and, in the case of the universe, from the original 'big bang' explosion

that created it), matter integrates; it develops order. <u>Thus, subject to the influence of Negative Entropy, the 94 elements from which our world is built develop ever larger and more stable wholes.</u>

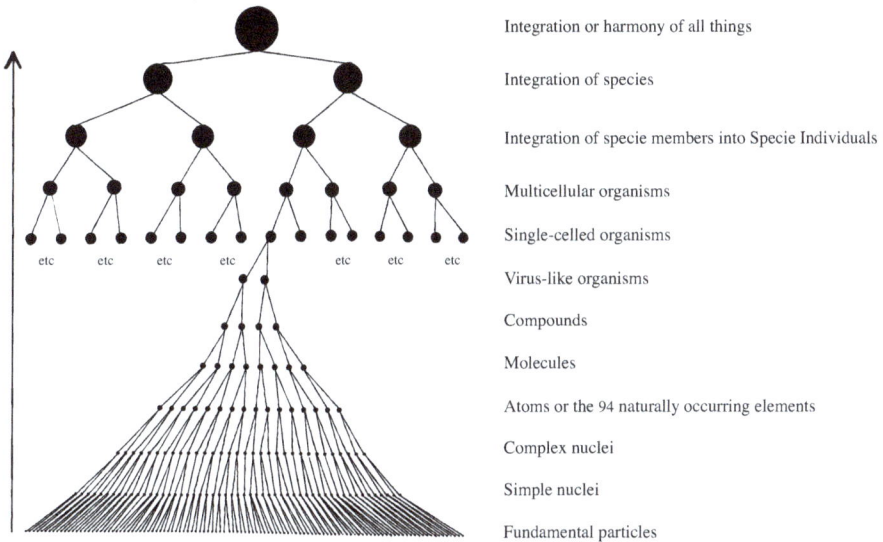

Integration or harmony of all things

Integration of species

Integration of specie members into Specie Individuals

Multicellular organisms

Single-celled organisms

Virus-like organisms

Compounds

Molecules

Atoms or the 94 naturally occurring elements

Complex nuclei

Simple nuclei

Fundamental particles

Chart showing the ordered integration of matter on Earth
(a similar chart appears in Arthur Koestler's book *Janus: A Summing Up*).

[60] Significantly, the behaviour required for these ordered arrangements of matter or wholes to stay together is selflessness — because selflessness means being considerate of the welfare of the larger whole, it means being integrative, while selfishness is divisive or disintegrative. Selflessness is in fact the theme of existence, the glue that holds wholes together. It is also what we mean by the word 'love', with the old Christian word for love being '**caritas**', meaning charity or giving or selflessness (see Col. 3:14, 1 Cor. 13:1-13, 10:24 & John 15:13). So 'love' is cooperative selflessness — and not just selflessness but *unconditional selflessness*, the capacity, if called upon, to make a full, self-sacrificing commitment to the maintenance of the larger whole.

[61] But, as pointed out, in light of our *divisive* competitive, selfish and aggressive behaviour, we have found this truth of the selflessness-dependent, integrative, cooperative, selfless and loving theme or meaning or nature of existence unbearably condemning. And being so unbearably condemning, the ways we coped with the truth of Integrative Meaning were firstly to deify it, make it 'God' — an ethereal concept we revered but

claimed had no material relationship to us; and secondly, in the case of human-condition-avoiding Reductionist, Mechanistic science, to simply deny its existence by maintaining that there is no direction or purpose or meaning to existence, and that change is random.

[62] Yes, since the terms 'Holism' and 'Teleology' recognise this truth of Integrative Meaning—'Holism' being defined as **'the tendency in nature to form wholes'** (*Concise Oxford Dictionary*, 5th edn, 1964), and 'Teleology' being defined as **'the belief that purpose and design are a part of nature'** (*Macquarie Dictionary*, 3rd edn, 1998)—human-condition-avoiding Reductionist, Mechanistic scientists have been deniers of the truth of Holism and Teleology.

[63] So the Integrative Meaning of life requires selfless consideration for the maintenance of larger wholes. Therefore, as initially stated, to be thinking in an Integrative-Meaning-compliant, truthful, effective, meaningful, 'conscious' way depended on being able to behave selflessly. Outside of the nurturing process that, as was just explained, allowed our ancestors to become selflessly behaved, selfless behaviour is blocked from developing in other animals because they have to selfishly ensure their genes are passed on, so as soon as they begin to think truthfully in an Integrative-Meaning-compliant way and so behave selflessly, natural selection blocks that truthful, effective, conscious thinking from developing. Other species that are competitively, selfishly and aggressively behaved and not cooperatively, selflessly and lovingly behaved, have minds that are blocked from thinking truthfully, effectively and consciously—just as we human-condition-avoiding humans, who have been unable to admit the truth of the Integrative Meaning of existence, and the truth of the integrative, cooperative, selfless and loving behaviour of our distant ancestors, have minds that are blocked from being able to think truthfully and effectively, and, for example, can't find all the insights that are being provided in this human-condition-confronting truthful presentation. Alienation blocks truthful thinking, it is a form of unconsciousness. So the reason the bonobos mentioned earlier in the quote in paragraph 43 exhibit a degree of consciousness is because they are becoming cooperatively behaved. Since this is an extremely brief description of how we became fully conscious when other species haven't, it will likely be necessary to read chapter 7 of *FREEDOM* or Freedom Essay 24 to fully understand, evaluate and appreciate the explanation.

[64] Again, I should emphasise that the reason you haven't heard of the truth of Integrative Meaning, and also of this explanation for the origin of our conscious mind, is because their acknowledgment depended on being

able to explain why, in the case of Integrative Meaning, we humans have not been cooperatively behaved, and, in the case of being able to explain how we became conscious, admit that our distant ape ancestors were co-operative and loving, which are both truths we couldn't admit while we couldn't explain our psychologically distressed, upset, angry, egocentric and alienated human condition—which is the explanation that is going to be presented. So many insights into the nature of our world and our place in it depend on being able to explain the human condition—which is why so many new concepts are having to be introduced in this book.

[65] To be able to 'open the door' to understanding ourselves we first had to solve the human condition. Without the explanation of the human condition, all that human-condition-avoiding Reductionist Mechanistic Scientists have been able to come up with to supposedly 'explain' human behaviour are an endless stream of lies—such as that our ancestors were brutal savages, which was mentioned earlier, or the just-mentioned lie that there has been no meaning or direction in our lives and that change is random, and, as we will see, the lies go on and on.

[66] Indeed, if we imagine an enormous building with lots of hallways and doorways, and any doorway that when opened revealed a truth that reminded us of the, soon to be explained, suicidally depressing truth that we had seemingly committed the most obscene crime of destroying our original all-harmonious, all-loving and all-sensitive, innocent, instinctive self or soul's world, which, as we will see, is basically any condemning, guilt-implying criticising truth, we would slam that door shut and never go into that room and turn its light on. The result is a huge building with only a very few rooms that have been opened and their lights turned on—because there are almost no truths about us humans that, up until now, haven't brought the unbearably depressing issue of our soul-corrupted human condition into focus. So what we metaphorically have is a huge building—basically the whole world—that is in almost total darkness! Such has been the darkness, the lack of understanding, in our world while we couldn't explain the human condition—and such is the light-filled brightness of relief and happiness that comes to the human race now that we can turn all those lights on, as is going to be done! I should mention that later in paragraphs 155-159 I include a very similar metaphor given by Plato of our denial of the human condition being like living in darkness; in Plato's case he used the analogy of the human race being imprisoned in a dark cave.

[67] As will now be explained, we humans have had to suffer from the agony and horror of the human condition—live imprisoned in Plato's dark cave—for some 2 million years, which is when the development of a fully conscious mind in our forebears occurred.

Part 1.6 The 'Instinct vs Intellect' explanation: how a clash between our cooperative, selfless and loving moral instincts and the emergence of our fully conscious mind caused our psychologically distressed, angry, egocentric and alienated human condition —the explanation of which relieves and ends that condition forever!

[68] What will now be explained is how an inevitable clash between our cooperative, selfless and loving moral instincts and our fully conscious, self-managing mind caused our psychologically distressed, angry, egocentric and alienated human condition to emerge.

[69] Firstly, as just mentioned, the development of a fully conscious mind in our forebears happened some 2 million years ago—that being the approximate time our large 'association cortex' (thinking) brain appears in the fossil record, which you can see some evidence of in this sequence of fossil skulls of our ancestors.

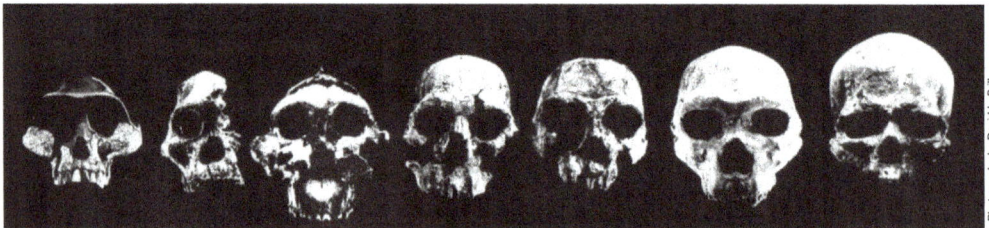

Photographs by David L. Brill

Australopithecus afarensis	*Australopithecus africanus*	*Australopithecus boisei*		*Homo habilis*	*Homo erectus*	*Homo sapiens*	*Homo sapiens sapiens*
Fossil evidence from 3.9 to 3 million years ago	3.3 to 2.1 m y a	2.3 to 1.2 m y a	Human Condition Fully Emerges Here	2.4 to 1.4 m y a	1.9 to 0.1 m y a	0.5 to 0.1 m y a	0.2 m y a to now
Brain Volume 400 cc average	450 cc	530 cc		650 cc	900-1100 cc	1350 cc	1400 cc
Early Happy Childman	Middle Demonstrative Childman	Late Naughty Childman		Distressed Adolescentman	Adventurous Adolescentman	Angry Adolescentman	Pseudo idealistic and Hollow Adolescentman

Humanity's stages of maturation, see chapter 8 of *FREEDOM*.
(Note, our large brain appeared some 2 mya.)

[70] As was explained in the previous section about consciousness, nerves were originally developed for the coordination of movement in animals. However, once developed, their capacity to store impressions—which is what we refer to as 'memory'—led to the ability to sufficiently associate information to reason how experiences are related, learn to understand and become CONSCIOUS of, or aware of, or intelligent about, the relationship between events that occur through time. Thus consciousness means being sufficiently aware of how experiences are related to attempt to manage change from a basis of understanding.

[71] What is so significant is that once our nerve-based learning system became sufficiently developed for us to become fully conscious some 2 million years ago and able to effectively manage events, that conscious intellect was then in a position to wrest control of our lives from our gene-based learning system's instincts, which up until then had been in charge. Basically, once our self-adjusting intellect emerged it was capable of taking over the management of our lives from the instinctive orientations we had acquired through the natural selection of genetic traits that adapted us to our environment. HOWEVER, it was at this juncture, when our conscious intellect challenged our instincts for control, that a terrible battle broke out between our instincts and intellect, the effect of which was the emergence of our psychologically distressed, angry, egocentric and alienated, human condition.

[72] So, when our conscious intellect emerged it was neither suitable nor sustainable for it to be *orientated* by instincts—it *had to* find *understanding* to operate effectively and fulfil its great potential to manage life. The obvious problem was that when our intellect began to exert itself and experiment in the management of life from a basis of understanding, it in effect challenged the role of our already established instinctual self, which naturally caused a battle to break out between our instinctive self and newer conscious self.

[73] Our intellect began to experiment in understanding as the only means of discovering the correct and incorrect understandings for managing existence, but the instincts—being in effect 'unaware' or 'ignorant' of the intellect's need to carry out these experiments—'opposed' any understanding-produced deviations from the established instinctive orientations: they 'criticised' and 'tried to stop' the conscious mind's necessary search for knowledge. To illustrate the situation, imagine what would

happen if we put a fully conscious mind on the head of a migrating bird. The bird is following an instinctive flight path acquired over thousands of generations of natural selection, but it now has a conscious mind that needs to *understand* how to behave, and the only way it can acquire that understanding is by experimenting in understanding—for example, thinking, 'I'll fly down and explore that island.' But such a deviation from the migratory flight path would naturally result in the instincts resisting the deviation, leaving the conscious intellect in a serious dilemma: if it obeys its instincts it will not feel 'criticised' by its instincts but nor will it find knowledge. Obviously, the intellect could not afford to give in to the instincts, and unable to understand and thus explain why its experiments in self-adjustment were necessary, the conscious intellect had no way of refuting the implicit criticism from those rigid, fixed, inflexible, 'dictatorial' instincts, even though it felt that criticism wasn't justified. In the case of humans, the species that had become fully conscious, until our conscious mind found the redeeming understanding of *why* it had to defy the instincts (namely the scientific understanding that is being presented here of this difference in the way genes and nerves process information, that genes are an orientating learning system while nerves, when much developed to the point of having become fully conscious, are an insightful learning system), our intellect was left having to endure what was in effect condemning criticism from our instincts, which understandably left us, our conscious thinking self, no choice but to defy that condemning opposition from our instincts. The only forms of defiance available to our conscious intellect were to attack our instincts' unjust criticism, attempt to prove our instincts' unjust criticism wrong, and try to deny or block from our mind the instincts' unjust criticism.

[74] Our mind became upset by that criticism that it felt was not justified but couldn't explain why it wasn't justified, and as a result it became defensively angry, egocentric and alienated. Our **'conscious thinking self'**, which is the *Concise Oxford Dictionary* definition of **'ego'** (5th edn, 1964), became 'centred' or focused on the need to justify itself. We became ego-centric, self-centred—selfishly preoccupied trying to prove we are good and not bad, competing for opportunities to prove our worth, and feeling the need to attack any criticism. We unavoidably became selfish, competitive and aggressive.

Part 1.7 What the human condition *really* is

[75]What now needs to be pointed out, AND THIS IS EXTREMELY IMPORTANT, is that *our* conscious thinking self wasn't in defiance of just any dictatorial instincts, which would have been upsetting enough, *we were in defiance of our particular cooperative, selfless and loving moral instincts*. What this means is that when we became competitive, selfish and aggressive, that response was *extremely* offensive to our particular cooperative, selfless and loving moral instincts. We, our conscious mind, had turned utopia into dystopia, destroyed paradise!

[76]Yes, as has been explained, we humans once lived instinctively in a nurtured-with-unconditional-love, cooperative, selfless and loving innocent state. It follows that since our particular instinctive orientation is to behaving cooperatively, selflessly and lovingly, the competitive, selfish and aggressive behaviour that resulted from becoming angry, egocentric and alienated only compounded the condemnation from our instincts, making us, our conscious thinking mind, *extremely*, *excruciatingly* angry, egocentric and alienated. And while we blocked out of our mind any criticising truth and as a result became separated from or alienated from so many truths, the most condemning of all truths that we were blocking out and repressing and alienating ourselves (our conscious thinking self) from was our condemning moral instinctive self or soul—we were having to destroy the most wonderful part of ourselves; but we had no choice because we were being so unjustly condemned by our soul. So our mind has been EXTREMELY angry, egocentric and alienated—which means we now have the reason for our capacity for shocking acts of violence, depravity, indifference and cruelty! We humans have endured intolerable guilt and shame for not only having defied our instincts, but having defied them in a way that drew *tremendous*, *unbearable* condemnation from our particular cooperative, selfless and loving moral instincts!

[77]So, as I point out in chapter 3:5 of *FREEDOM*, we (our conscious mind) have suffered from a 'double whammy' of condemnation for being competitive, selfish and aggressive. In fact, we have suffered from a 'triple whammy' of condemnation because our competitive, selfish and aggressive behaviour was *divisive* behaviour, so we were behaving entirely inconsistently with the Integrative Meaning of existence, we were seemingly out of step with creation—we were (as was explained earlier

in paragraph 61 when 'God' was explained) offending 'God'! Of course, our mind hasn't been able to comprehend this thought journey, it has just felt extremely upset, which we can now fully unpack and understand the reasons for.

[78]To repeat what happened: our species' original cooperative, selfless and loving moral instinctive self, the 'voice' or expression of which is our 'conscience', is what we have historically been referring to as our 'soul' or 'psyche', and it was this cooperative, selfless and loving instinctive part of ourselves that resisted, in effect condemned, our conscious mind's experiments in self-management. This 'condemnation' naturally upset our conscious mind, causing it to become defensively angry, egocentric and alienated, which resulted in it becoming competitively, selfishly and aggressively behaved. Since our particular instinctive self or soul was orientated to behaving cooperatively, selflessly and lovingly, it was *extremely* offended by this competitive, selfish and aggressive response. Our conscious self defied our instincts by experimenting in understanding, then further offended our integrative, cooperative, selfless and loving instincts by behaving in a divisive, out-of-sync-with-the-meaning-of-existence 'unGodly', competitive, selfish and aggressive way. So our instinctive soul was extremely offended and distressed by our conscious intellect having corrupted or destroyed its all-harmonious, all-loving and all-sensitive world—it was deeply hurt by the brutal, seemingly unGodly, competitive, selfish and aggressive behaviour that had appeared. But from our conscious mind's point of view, it was being made to suffer a double and triple whammy of condemnation, which naturally upset it a great deal, causing it to become *extremely* determined to try to attack, prove wrong, and deny this further condemnation. And the best way for our mind to deny that condemnation was for it to block out from its thoughts, and by so doing repress the presence of our extremely condemning instinctive self or soul—with the result being the extremely upset state of our mind or nerves or neurons, which is our mind's neurotic neurosis, and our mind's extreme repression of our instinctive soul or psychotic psychosis. We, our conscious mind, became neurotic and psychotic—basically, *mind-mad and soul-dead!*

[79] So, overall, OUR INSTINCTIVE SOUL WAS *EXTREMELY* OFFENDED, DISTRESSED AND HURT, AND OUR CONSCIOUS MIND WAS *EXTREMELY* UPSET, which was a horrific state or condition, OUR HUMAN CONDITION, that we humans have had to endure!

[80] Looking at the etymology of the terms being used, our 'conscience' is defined by the *Concise Oxford Dictionary* as our **'moral sense of right and wrong'**, and our 'soul' as the **'moral and emotional part of man'**, and as the **'animating or essential part'** of ourselves (5th edn, 1964), while the *Penguin Dictionary of Psychology*'s entry for 'psyche' reads: **'The oldest and most general use of this term is by the early Greeks, who envisioned the psyche as the soul or the very essence of life.'** Indeed, as the **'early Greek'** philosopher Plato wrote about our innate, ideal-or-Godly-behaviour-expecting moral nature, we humans have **'knowledge, both before and at the moment of birth…of all absolute standards…**[of] **beauty, goodness, uprightness, holiness…our souls exist before our birth…**[our] **soul resembles the divine'** (*Phaedo*, c.360 BC; tr. H. Tredennick, 1954, 65-80). So the word 'psychology' literally means **'science or knowledge of soul'**, and 'psychosis' literally means **'soul-illness'**, and 'psychiatry' literally means **'soul-healing'** (derived as they are from *psyche* meaning **'soul'**, *ology* meaning the **'branch of knowledge or science'**, *osis* meaning **'abnormal state or condition'** (*Dictionary.com*), and *iatreia* meaning **'healing'** (*The Encyclopedic World Dictionary*)). Yes, the 'double and triple whammy' finally explains our *extremely* soul-repressed psychosis or **'soul-illness'**, and our extremely mind-upset neurosis or neuron or nerve or **'intellect-illness'**. And again, as I said in paragraph 23, according to the *Cambridge Dictionary* 'psychological' means **'relating to the human mind and feelings'**, which we can now understand is our mind's upset neurosis and soul-repressed psychosis, WHICH IS ALL WHAT OUR HUMAN CONDITION *REALLY* IS!

[81] It is obviously very important to properly understand this situation that created our mind-mad and soul-dead human condition. When our fully conscious mind emerged some 2 million years ago its experiments in understanding (in the bird analogy, the bird's deviation from its migratory path to explore an island) were in effect condemned by our instincts, and that condemnation upset our conscious mind, causing it to become

defensively angry, egocentric and alienated, the effect of which was that we, our conscious thinking self, became competitive, selfish and aggressively behaved. Since our instinctive orientation was to be cooperative, selfless and loving, this response from our conscious mind was doubly OFFENSIVE AND DISTRESSING to our moral instinctive self or SOUL, making us, our conscious MIND, *extremely* UPSET with defensive anger, egocentricity and alienation. There was also the triple whammy of criticism because our divisive behaviour was inconsistent with the Godly Integrative Meaning of existence. No wonder we, us conscious thinking humans, have been so *extremely* psychotic or soul-repressed, and so *extremely* neurotic or mind-upset!

[82] So we defied our instincts, we offended our moral conscience, and we also insulted the very meaning of existence or 'God'! We humans could hardly have become more guilt-ridden when we became conscious and set out in search of understanding; ultimately for understanding of why, through undertaking that search, we destroyed or corrupted the all-harmonious, all-loving and all-sensitive world of our soul. And all this shame and guilt, which we can now understand was completely unjustified, made us excruciatingly guilt-ridden and overwhelmed with depression—and furiously angry, which, as I said, is why we have been capable of horrific acts of brutality, barbarism, sadism and cruelty. While we have tried to stay away from the deep depression by constantly maintaining a positive, buoyant, laughing, happy, 'we-are-absolutely-fine' persona, and tried to restrain and conceal the anger within us—be 'sociable' and 'civilised' as we say—there is a deep underlying depression which we are always trying to stay away from, and, in truth as well, a volcanic anger within us that we're always trying to restrain. So thank goodness we can now at last compassionately understand that we humans are fundamentally good and not bad, and by so doing lift forever an underlying dark shadowy world of depression within us and be truly happy instead of artificially happy—and also end forever our deep seated anger and be truly kind and loving towards ourselves, others and the whole world instead of being artificially restrained and civilised! End forever all the psychological upset neurosis and psychosis in the world; let the sunshine of understanding in and flood the dark world we have been living in with true and real peace, love and happiness!

J Griffith, M Rowell and G Salter © 2009 Fedmex Pty Ltd

[83] Yes, most wonderfully, since we became defensively angry, ego-centric and alienated because we couldn't explain why we were defying our instincts, now that we *can* explain why, those defensive behaviours are no longer needed and can end, and by doing so all our psychosis and neurosis will disappear—which is a dreamed-of transformation that is going to be explained more fully later in Part 2.9!

[84] As I said in the Brief Summary, it is science that has made this redeeming, reconciling and rehabilitating explanation for why we be-came angry, egocentric and alienated possible because it wasn't until science revealed the difference between the gene-based and nerve-based learning systems—which is that genes can orientate but nerves need to understand—that we were finally in a position to explain the good reason for our mind's extremely upset angry, egocentric and alienated neurosis and its extremely soul-repressing psychosis. And I should say, as I men-tioned in paragraph 7, now that it has been explained, this Instinct versus Intellect, soul versus mind, genes versus nerves, two-main-influences-on-our-behaviour-that-when-you-think-about-the-nature-of-them-have-to-have-clashed, explanation for the human condition is a reasonably obvious, straightforward and simple explanation—which bears out biologist Allan Savory's observation that **'whenever there has been a major insoluble problem for mankind, the answer, when finally found, has always been very simple'** (*Holistic Resource Management*, 1988, 1st edition, p.3).

Adam and Eve, by
Lucas Cranach the Elder (1526)

Adam and Eve cast out of Paradise, from
Old Testament Stories pub. Society for
Promoting Christian Knowledge, London (c.1880)

[85] The Biblical story of Adam and Eve in the Garden of Eden that Moses wrote so long ago in about 1,500 BC actually perfectly describes the soul-offending and mind-upsetting clash that emerged between our fixed, dictatorial, moral instincts and self-managing conscious intellect. It says Adam and Eve (we humans) took the **'fruit'** (Genesis 3:3) **'from the tree of knowledge'** (Gen. 2:9, 17) and were **'disobedient'** (the term widely used in descriptions of Gen. 3); in other words, we developed a conscious mind and free will. But in that *pre-scientific* story it says Adam and Eve then became **'evil'** (Gen. 3:22) perpetrators of **'sin'** (Gen. 4:7) because, as we can now understand, they had become angry, egocentric and alienated, and as a result Moses said Adam and Eve (we humans) were **'banished…from the Garden of Eden'** (Gen. 3:23) state of cooperative, selfless and loving innocence.

[86] Not knowing how naturally selected instincts differ from cause and effect-operating consciousness, this story of Adam and Eve becoming conscious could only conclude that the angry, egocentric and alienated, soul-corrupted condition that emerged when we became conscious was a bad, evil, sinful state, but this *scientific* presentation says, 'No, no, that pre-scientific story got it wrong'. Adam and Eve, we humans, are actually not just good but the heroes of the whole story of life on Earth—because surely the conscious mind is nature's greatest invention and to be given the task of searching for understanding while the whole world's condemning you was the hardest and toughest of tasks—because that condemnation *was* universal. All of nature—the rain, the clouds, the trees, and other

animals—are all associated with, and are thus, as it were, friends of our original innocent instinctive self or soul that was condemning us no-longer-innocent, seemingly bad, awful conscious humans! The whole world, in fact the whole of the integrated universe, in effect, ganged up on Adam and Eve, i.e. on us conscious humans—and yet all the time we were good and not bad but we couldn't explain why, but now at last through the benefit of science, we can. We can now explain and understand that we conscious humans are immensely heroic, and not villains after all!

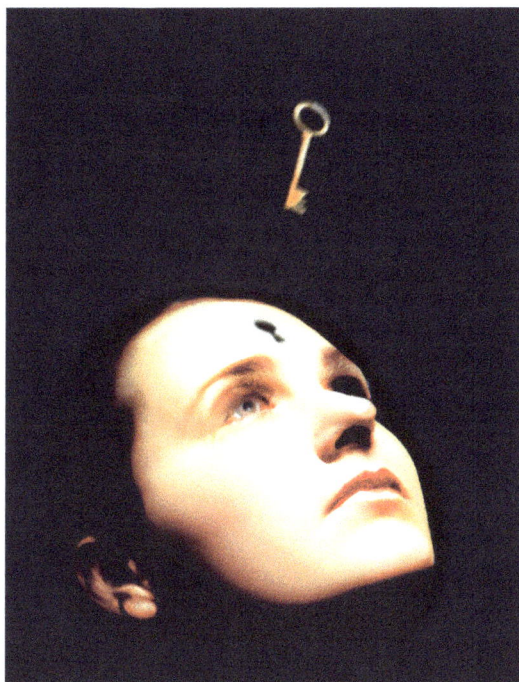

© 2023 Fedmex Pty Ltd

[87] Yes, this ability to understand and know there was a good reason why the human race became so psychologically distressed—so mind-upset or neurotic with anger, egocentricity and alienation, and so soul-repressed or psychotic—is the key, relieving understanding of ourselves that we have been in search of ever since we became conscious some 2 million years ago and our horrifically neurotic and psychotic human condition emerged. The psychoanalyst Carl Jung famously said that **'wholeness for humans depends on the ability to own our own shadow'**, but we have never been able to **'own'** the **'shadow'** of our species' horrifically 2-million-year mind-upset and soul-repressed condition and become **'whole'**; 'explain' to our original instinctive self or soul that we, our fully conscious thinking

self, is good and not bad and by so doing reconcile and heal our split selves—but now at last we can. As it says in the Bible, **'the truth will set you free'** (John 8:32), and that is what has happened: the redeeming truth about our divisive mind-enraged and soul-destroyed behaviour has set us free from the insecurity and horror of that condition! We have the key understanding that takes the pain out of our brains! One of the founders of the 1980s pseudo idealistic New Age Movement (see three paragraphs below), the author Marilyn Ferguson, was actually hoping for the psychologically redeeming understanding of the human condition that we now have when she wrote: **'Maybe** [the Jesuit priest, scientist and philosopher] **Teilhard de Chardin was right; maybe we are moving toward an omega point** [a final genuine unification/individuation of our split selves]**...Maybe...we can finally resolve the planet's inner conflict between its neurotic self (which we've created and which is unreal) and its real self. Our** [original all-sensitive and loving instinctive] **real self knows how to commune, how to create...From everything I've seen people really urgently want the kind of new beginning...** [that I am] **talking about** [where humans will live in]**...cooperation instead of competition'** (*New Age* mag. Aug. 1982; see www.wtmsources.com/174).

[88] Of course, the Biblical story of Adam and Eve in the Garden of Eden is far from the only recognition from ancient times of the elements of our moral instincts and our conscious intellect being involved in producing our distressed human condition. Indeed, as the journalist and author Richard Heinberg summarised in his 1990 book *Memories & Visions of Paradise*, **'Every religion begins with the recognition that human consciousness has been separated from the divine Source, that a former sense of oneness...has been lost... everywhere in religion and myth there is an acknowledgment that we have departed from an original...innocence and can return to it only through the resolution of some profound inner discord...the cause of the Fall is described variously as disobedience, as the eating of a forbidden fruit** [from the tree of knowledge]**, and as spiritual amnesia** [forgetting, blocking out, denial, alienation, which is our psychosis]**'** (pp.81-82 of 282). So all our religions and most of our mythologies have recognised the basic conflict within us—that the emergence of **'consciousness'** caused our **'Fall'** from **'innocence'**. The quotes from Hesiod and Plato that were included earlier in paragraphs 49-50 are examples of mythologies that recognised our species' time in innocence before we became conscious. What was needed in all these religious and mythological recognitions of our species' pre-conscious time in innocence was the redeeming good reason for WHY **'a former sense of oneness...has been lost'** and **'we...departed**

from an original…innocence', and HOW we 'can [now] return to it…through the resolution of some profound inner discord', which is the 'instincts can orientate but nerves need to understand' explanation that we now have.

[89] So our competitive, selfish and aggressive behaviour is not due to savage, must-reproduce-our-genes instincts but to an *extremely* upset conscious mind and an *extremely* distressed soul. Basically, becoming upset, angry, egocentric and alienated, and, as a result, competitively, selfishly and aggressively behaved, and as a result of that becoming neurotic and psychotic, has been the price we conscious humans had to pay for our heroic search for knowledge, ultimately for self-knowledge, the redeeming biological explanation for why this all happened, which is the explanation that has just been provided. In the words of the song *The Impossible Dream* from the musical the *Man of La Mancha*, we had to be prepared to **'march into hell for a heavenly cause'** (lyrics by Joe Darion, 1965). We had to lose ourselves to find ourselves; we had to suffer becoming angry, egocentric and alienated and as a result competitively, selfishly and aggressively behaved and neurotic and psychotic until we found sufficient knowledge to explain why this all happened.

[90] We humans have been trapped in a horrifically paradoxical situation where to free ourselves from being angry, egocentric and alienated, and competitive, selfish and aggressive, we had to continue behaving like that until we found the explanation for why we were behaving like that! Only having the freedom to be competitive, selfish and aggressive could lead to the finding of the understanding of ourselves that would enable us to stop having to be competitive, selfish and aggressive! As I explain in my book *Death by Dogma*, and in many of my presentations, the dogmatic insistence by the left-wing in politics that everyone stop being competitive, selfish and aggressive actually blocked the heroic competition-selfishness-and-aggression-producing search for knowledge needed to find the redeeming understanding for why we were being competitive, selfish and aggressive that was required to end the need to be defensively competitive, selfish and aggressive! It is actually the right-wing's support of the soul-destroying competition-selfishness-and-aggression-producing search for knowledge that held the moral high ground, *not* the left-wing's dogmatic, pseudo idealistic insistence on cooperative, selfless and loving behaviour and its dogmatic, pseudo idealistic oppression of competitive, selfish and aggressive behaviour. Dogma is not the cure, it's the poison. The Left are regressive, *not* progressive as they delude themselves they

are in order to feel good about themselves—artificially relieve themselves of feeling bad about their and the world's corrupted state of the human condition. Thank goodness, the redeeming explanation for our soul-corrupted angry, egocentric and alienated human condition has now been found, so the corrupting anger, egocentricity and alienation-producing search for that redeeming self-knowledge is over, and the whole torturous business of politics comes to an end!

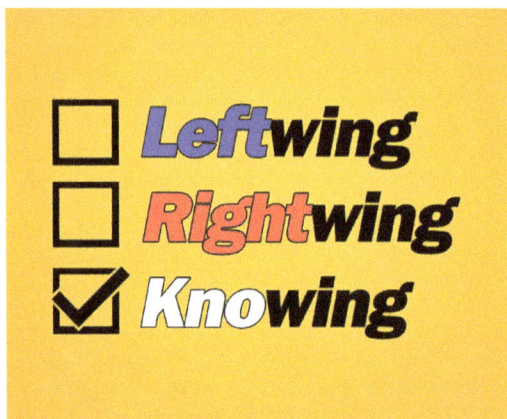

[91] I should explain that in *THE Interview* and in a number of my shorter presentations where I have needed to keep the explanation of the human condition brief, I limited the explanation of the human condition to the clash between our instincts and intellect and didn't go into the 'double and triple whammy' effects on our moral instincts when we became angry, egocentric and alienated. As will likely be very apparent by now, even in this somewhat longer presentation, there are so many new truthful-human-condition-confronting-not-dishonest-human-condition-avoiding insights to have to be introduced and grasped that presenting the 'double and triple whammy' effect would be too much to include in a shorter version like *THE Interview*. While I didn't explain the 'double and triple whammy' effects in those shorter presentations, they are inferred because I do point out in them that we have cooperative, selfless and loving moral instincts, and that we suffered from extreme guilt and shame.

[92] IN SUMMARY, WHAT HAS NOW BEEN EXPLAINED FINALLY ENABLES US TO APPRECIATE WHAT THE ISSUE OF THE HUMAN CONDITION *REALLY* REFERS TO. It is the horrific shame and guilt we humans have suffered from for not only having defied our instinctive

orientations to the world but having defied them in a competitive, selfish and aggressive way that drew *tremendous* condemnation from our particular cooperative, selfless and loving, Integrative-Meaning-complying moral instincts. Basically, the human condition is the unbearable shame and guilt of having corrupted/destroyed the original all-harmonious, all-loving and all-sensitive, innocent instinctive world of our soul!

[93] This insufferable guilt and shame we conscious thinking humans have experienced for destroying the magic world of our soul has, in fact, been so unbearably depressing, indeed so suicidally depressing, that the only sensible way that we have been able to cope with our seeming horrific crime while we couldn't explain it was to deny that we ever lived in a cooperative, selfless and loving, innocent state, and, by extension, by denying that we suffer from any soul-repressed, psychotic and mind-upset, neurotic, mad, deranged, profoundly-disturbed state, when we actually certainly do! The truth is we humans developed an *extreme* 'psychosis' or **'soul-illness'**, and also an *extremely* upset mind, a '[neurosis]' or neuron or nerve or **'intellect-illness'**. We have a *very* sick mind and *very* sick soul!

[94] Indeed, it makes sense that this practice of denial of our extreme soul-corrupted psychosis and extreme mind-distressed neurosis has been *so* important to us that it has been the dominant feature of our behaviour. We have been 'A Species In Denial', as I titled my 2003 book that became a bestseller in Australia and New Zealand. The polymath Johann Wolfgang von Goethe was truthfully recognising the extent of our immensely soul-repressed (psychotic) and mind-upset (neurotic) state when he famously said, **'We do not have to visit a madhouse to find disordered minds; our planet is the mental institution of the universe'**, as was the playwright George Bernard Shaw when he famously said, **'The longer I live, the more convinced am I that this planet is used by other planets as a lunatic asylum'**!

[95] Given this desperate need to deny we ever lived in a cooperative, selfless and loving innocent state while we couldn't explain our seemingly evil instinctive-self-or-soul-corrupted condition, we can now appreciate that even though the must-reproduce-our-genes-like-other-animals, savage instincts excuse was an obviously false excuse it was an absolutely brilliantly helpful excuse because instead of our instincts being all-loving and thus unbearably condemning of our present non-loving state, they were made out to be competitive, selfish and aggressive must-reproduce-your-genes instincts like other animals have; *and*, instead of our conscious mind being the instinct-defying cause of our corruption, which it actually was, it

was made out to be the blameless mediating 'hero' that had to step in and try to control those supposed savage instincts within us! What complete reverse-of-the-truth lies, but so relieving to have while we searched for the real explanation for our psychologically distressed condition! And I should also mention our evasive tactic where those who dared to admit the truth of our cooperative, selfless and loving past, like Hesiod and Plato (in paragraphs 49-50), were dismissed as deluded romantics, and the whole idea of an innocent, Edenic past was said to be nothing more than a nostalgia for the security and maternal warmth of infancy; that it was **'never an historical state'** as the human-condition-avoiding Jungian psychologist Erich Neumann said in his book *The Origins and History of Consciousness* (1949, p.15 of 493). What another brilliant but outrageous reverse-of-the-truth lie! As I said earlier in paragraph 65, lying has absolutely been a human specialty—well, most unfortunately, it had to be, but thank goodness all that horrific lying now stops! As I said, we now let the light of honesty in to replace all that lying darkness. We, the human race, are coming home to peace, truth and happiness at last!

[96] This necessary denial of our soul-corrupted condition while we couldn't explain it is why all these truths that are being revealed have been hidden from us—such as the truth of the Integrative Meaning of existence; the truth that our savage, must-reproduce-our-genes excuse for our divisive behaviour is a false excuse; the truth that we have cooperative, selfless and loving moral instincts; the truth that nurturing created our moral soul and enabled us to develop a fully conscious mind—and, as will be pointed out in the next few paragraphs, the truth about why we have lived such an *extremely* fearful, psychologically insecure, superficial and artificial existence; and about how finally having the redeeming Instinct vs Intellect good reason for our soul-corrupted condition lifts the great so-called 'burden of guilt' from our species' conscious mind, and liberates the entire human race from that tortured condition, making possible a whole new, neurologically relieved and psychologically rehabilitated, sane, peaceful and loving transformed world—hence why we are called the World Transformation Movement!

[97] Even though we humans have had to resign ourselves to living in determined denial of the truth that our species did once live in a co-operative, selfless and loving, innocent state, and in determined denial of the issue it raises of why we committed the seemingly horrific crime of

destroying such a peaceful and loving 'Garden of Eden' existence, these truths of our species' original innocence and the issue it raises of are we despicable monsters have been subconsciously preoccupying our mind every minute of our lives! We have, in truth, most exhaustingly spent every moment making ourselves feel good, keeping at bay the unbearably depressing thought that we are seemingly bad, unworthy, even evil beings. We have constantly sought reinforcement for ourselves, and always tried to stay positive to hold at bay the underlying depressing fear that we might indeed be dreadful mistakes on Earth!

[98] SO, THANK GOODNESS A MILLION TIMES OVER, ALL THAT INSECURE BEHAVIOUR CAN END. As the wonderfully transformed WTM founding member Tony Gowing has said, **'We don't have to be ashamed. We don't have to shake our fists at the heavens anymore and prove to everyone and everything that they have been wrong about us. First-principle science has proven that we are worthwhile; that we are gloriously heroic beings. It had to be the way it's been—there was no other way—but it's all over now. The relief of finally being able to understand floods through our whole being; the anger and frustration dissipates; all the bullshit, falseness and lies end. We can finally love ourselves and participate in the world instead of constantly fighting it. No longer preoccupied with proving our self-worth, we will finally have the room in ourselves to properly help others; to selflessly participate in stopping the suffering everywhere we look'** (Video/Freedom Essay 5). Yes, the human race returns home now to a state of human-condition-free sanity, peace, togetherness and happiness. As the poet T.S. Eliot so truthfully anticipated, **'We shall not cease from exploration and the end of all our exploring will be to arrive where we started and know the place for the first time'** (*Little Gidding*, 1942).

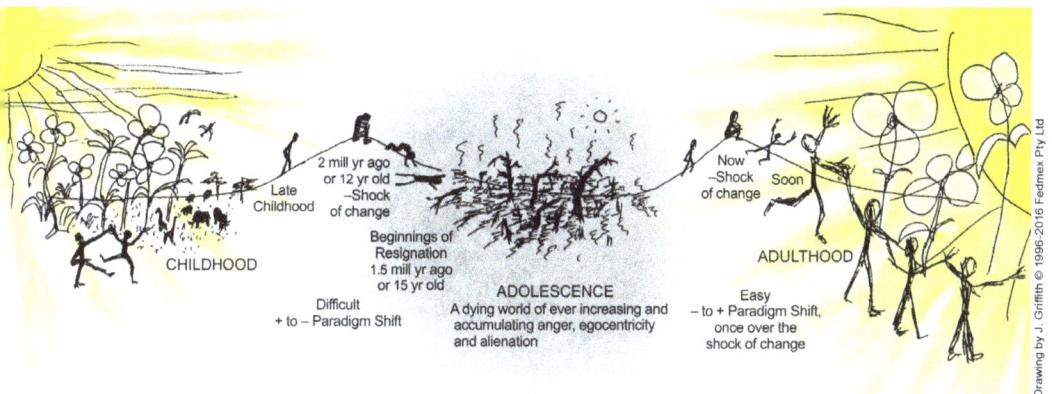

CHILDHOOD

Late Childhood

2 mill yr ago or 12 yr old −Shock of change

Beginnings of Resignation 1.5 mill yr ago or 15 yr old

Difficult + to − Paradigm Shift

ADOLESCENCE

A dying world of ever increasing and accumulating anger, egocentricity and alienation

Now −Shock of change

Soon

ADULTHOOD

Easy − to + Paradigm Shift, once over the shock of change

Drawing by J. Griffith © 1996-2016 Fedmex Pty Ltd

Humanity's Journey from Ignorance to Enlightenment, see chapter 8 of *FREEDOM*

[99] SO *THAT* IS WHAT THE HUMAN CONDITION *REALLY* IS: the torturous underlying neurological insecurity and psychological distress we fully conscious humans have suffered from for some 2 million years for having committed the seemingly inexcusable crime of corrupting our all-harmonious and all-loving and all-sensitive original instinctive self or soul!

[100] And yes, as explained more fully in Part 4 of *THE Interview*, and much more fully in the latter part of chapter 3 of *FREEDOM*, since we now have the redeeming and reconciling Instinct vs Intellect good reason for why we corrupted our original cooperative, selfless and loving instinctive self or soul, the immense underlying neurological insecurity and psychological distress about our corrupted condition comes to an end—and the complete neurological and psychological rehabilitation and transformation of our species can now take place! As Professor Harry Prosen, a former president of the Canadian Psychiatric Association, said, **'I have no doubt Jeremy Griffith's instinct vs intellect biological explanation of the human condition is the holy grail of insight we have sought for the psychological rehabilitation of the human race.'**

[101] So, we, the human race, has achieved Martin Luther King Jr's yearned-for **'dream'** to be **'Free at last! Free at last! Thank God Almighty, we are free at last!'** ('I Have A Dream' speech, 28 Aug. 1963).

Drawing by J. Griffith © 2010-2019 Fedmex Pty Ltd

Part 1.8 The truth of the 'Instinct vs Intellect' explanation is confirmed by its ability to make sense of all aspects of human life

[102] Before moving on to the 'Extended Description of the human condition' in Part 2, it should be pointed out that now that we can explain the human condition the human race goes through a massive paradigm shift from denial to honesty. All these insights that were hidden from us by our necessary practice of denial of any truths that brought the previously unbearably depressing issue of our corrupted condition into focus become possible—such as about Integrative Meaning; about our species' past state of innocence; about the falseness of the savage, must-reproduce-our-genes excuse for our divisive behaviour; about how nurturing gave us our moral soul and enabled us to develop a fully conscious mind; about why we have lived such an extremely fearful, psychologically insecure, superficial and artificial existence; and about how when the redeeming understanding of our soul-corrupted condition was finally found, it would lift the great, so-called, 'burden of guilt' from our species' conscious mind, and liberate the entire human race from that tortured condition, making possible a whole new, neurologically relieved and psychologically rehabilitated, sane, peaceful and loving transformed world. When understanding of the human condition is finally found, as it now has been, suddenly so many previously inaccessible explanations and insights about human life become accessible. Metaphorically, when the 'lights are switched on' in our dark, truth-avoiding, denial-saturated 'great building' referred to in paragraph 66, which is, in fact, our world, the 'light' of understanding streams in in all directions!

[103] And all these insights into human life—and many, many more of them will be revealed next in Part 2—together with the avalanche of explanations, answers and insights into human life that are presented in *FREEDOM*, confirm the truth of the Instinct vs Intellect explanation of the human condition. Indeed, the strength of a scientific theory is related to the diversity of phenomena it can explain—as the great physicist Stephen Hawking wrote, **'A theory is a good theory if it satisfies two requirements: It must accurately describe a large class of observations on the basis of a model that contains only a few arbitrary elements, and it must make definite predictions about the results of future observations'** (*A Brief History Of Time*, 1988).

[104] The evolutionary biologist Stephen J. Gould was applying a similar methodology to Hawking's when he argued that Charles Darwin's theory of Evolution by Natural Selection pointed to the coordination of so many pieces of evidence that only his theory could offer a conceivable explanation, and that in this way Natural Selection has, in effect, been proven. Gould wrote, **'Evolution is an inference from thousands of independent sources, the only conceptual structure that can make unified sense of all this disparate information'** ('Mr Sophia's Pony', *Leonardo's Mountain of Clams and the Diet of Worms*, 1988).

[105] The Instinct vs Intellect theory is very similar to Darwin's in its ability to **'make unified sense of'** so much **'disparate information'** and **'independent sources'** that it is clearly **'the only conceptual'** explanation, in this case, for the human condition.

[106] The following are just some of the **'disparate'** issues that the Instinct vs Intellect explanation **'make[s] unified sense of'**, all of which you can find in *FREEDOM* by looking through its index, or by searching the book electronically. At least the outline of most of these explanations is presented in this book.

- It finally explains and ends forever our neurotic mind-mad and psychotic soul-dead human condition, making possible the end of loneliness, depression and anxiety;
- It explains the origin of aggression and war amongst humans and brings them to an end;
- It explains and ends the need for materialism, and our need for a superficial, artificial, self-distracting way of living;
- It explains how humans acquired our altruistic 'soul' and its cooperation-demanding 'conscience';
- It explains how humans became fully conscious and why other animals have not;
- It explains the importance of nurturing in our species' development and in our individual lives, and the corrupting effects on children of mothers' inability now to nurture and of fathers' now extreme egocentricity;
- It explains why bonobos are the most cooperative of all extant apes;
- It provides a truthful explanation of the stages of maturation of infancy, childhood, adolescence and adulthood that both humanity and humans individually go through;
- It describes and explains the cause of teenage angst, which is the psychological act of Resignation;

- It explains the Negative Entropy-driven integrative, loving meaning of life, and why 'evolution' is in fact the purposeful process of ordering matter;
- It explains the extreme limitations of reductionist, mechanistic science and the immense benefits of holistic, teleological science;
- It relates and reconciles the studies of the sciences and of the humanities;
- It explains why and when humans learnt to walk upright, reduced the size of their canine teeth, lost their body hair, began tool use, began hunting and meat-eating, left Africa, and developed language;
- It reconciles science with religion, in the process explaining religion and all manner of religious metaphysics, including the concepts of God, the Trinity, the Virgin Mary, the resurrection, miracles, Judgment Day, the Battle of Armageddon, the stories of Adam and Eve including the Garden of Eden and taking the apple from the Tree of Knowledge, Noah's Ark and The Flood, Cain and Abel, David and Goliath, after-life, heaven and hell, good and evil, humanity's 'fall';
- It explains and demystifies Christ and the other great prophets;
- It deciphers humanity's legends and myths about the search for liberating understanding of the human condition such as the story of King Arthur's Knights of the Round Table's search for the 'Holy Grail', Jason and the Argonauts' search for the 'Golden Fleece', Ulysses's great adventure and return to Ithaca, *The Emperor's New Clothes*, *The Man from Snowy River* in Australia's mythology;
- It ends prejudice and inequality between individuals, sexes, ages, generations, races, cultures and civilisations;
- It explains and reconciles the left and right wings of politics, in the process ending the need for politics;
- It replaces the destructively deluded, pseudo idealistic transformation that the New Age, Peace, Green, Climate, Feminist, Indigenous, Animal Rights, Multicultural, Politically Correct, Postmodern, Critical Theory, 'Woke', Identity Politics, Great Reset, Socialist, Globalist movements were trying to introduce with the real psychologically healing transformation of the human condition;
- It explains and heals the ultimate source of all psychological disorders and the alarming growth in mental disease;

- It explains and heals the main underlying cause for human sickness;
- It explains sex, heterosexuality, homosexuality, love, physical beauty, the attraction of youth, romance, rape, envy and lust;
- It explains and brings an end to humans' propensity for paranoia and all manner of conspiracy theories, such as of lost civilisations with advanced technology;
- It explains human sensitivity and creativity, especially in art (including ancient cave painting) and music;
- It explains humour;
- It explains swearing;
- It explains near-death experiences;
- AND it saves the human race from self-destruction and brings peace and happiness to human life!

[107] Such is the power of thinking truthfully about the human condition!

[108] Accordingly, the Instinct vs Intellect treatise has received significant scientific evaluation, and, subsequent to that, support from the many extremely eminent truthful-thinking scientists, such as those who have provided the commendations that are listed at www.humancondition.com/reviews-commendations. For example, Professor Harry Prosen, a former President of the Canadian Psychiatric Society, wrote in his Introduction to *FREEDOM* that 'all the great theories I have encountered in my lifetime of studies of psychiatry can be accounted for under Jeremy Griffith's explanation of human origins and behavior'. And Professor Stuart Hurlbert, Emeritus Professor of biology at San Diego State University, wrote in 2024 that 'I am stunned & honored to have lived to see the coming of 'Darwin II'. I say this because after Darwin's theory of Natural Selection explained the variety of life, Jeremy Griffith has gone on to solve the other four main questions science had to answer about our world and place in it. They are: 1) the dilemma of the human condition, which his instinct vs intellect explanation in chapter 3 of his main, seminal book *FREEDOM* finally solves; 2) how we humans became fully conscious when other species haven't, which he answers in chapter 7; 3) the origins of humans' unique moral nature, which he answers in chapter 5, which it turns out American philosopher John Fiske had already explained in 1874 but mechanistic science had ignored; and 4) the truth of the Integrative Meaning of existence (which we have personified as 'God'), in chapter 4, which only a rare few thinkers in history have been able to recognize. And having been able to solve those primary issues he has, in chapter 8, using first principle and

fully accountable biological explanations been able to resolve all the secondary problems like: the polarized state of politics; the rift between men and women; the schism between science and religion; the conflict between individuals and between races (thus ending aggression and war at its source); and, above all, bring an end to the threat of terminal psychosis and our species' extinction! A truly phenomenal, beyond description, scientific achievement!' Additional commendations reflect the same extraordinary recognition. Professor Scott Churchill, a former Chair of Psychology at the University of Dallas, described *FREEDOM* as **'the book all humans need to read for our collective wellbeing'**, while former President of the Primate Society of Britain Professor David Chivers praised it as **'the necessary breakthrough in the critical issue of needing to understand ourselves.'** World-renowned psychologist Professor Mihaly Csikszentmihalyi suggested it **'might help bring about a paradigm shift in the self-image of humanity.'** Professor John Morton, zoologist at the University of Auckland, described my book *A Species In Denial* as **'a superb book…**[that] **brings out the truth of a new and wider frontier for humankind, a forward view of a world of humans no longer in naked competition amongst ourselves.'** And Templeton Prize winner Professor Charles Birch, Professor of Biology at the University of Sydney, praised the work for providing **'a genuinely original and inspiring way of understanding ourselves and our place in the universe.'** And even the just-mentioned Stephen Hawking, one of the most esteemed scientists of all time, was **'most interested in'** and **'impressed'** by the Instinct vs Intellect treatise (these responses to the treatise are among the commendations on the above 'Reviews & Commendations' page).

[109] So yes, as Professor Hurlbert wrote, the Instinct vs Intellect explanation of the human condition finally gives humanity the ability to **'bring an end to the threat of terminal psychosis and our species' extinction!'**

[110] With regard to the veracity of the Instinct vs Intellect treatise, it should also be pointed out that history teems with non-scientific mystical, superstitious and super-natural reasons for humans' often brutally aggressive and selfish nature or condition, such as that an evil force called 'Satan' emerged from some terrifying dark realm and enticed us humans into the clutches of evil and sin, condemning most of us to a dreadful purgatory in a fiery Hell. However, now that it has been explained, the rational, scientific Instinct vs Intellect explanation of the human condition is actually very obvious—because of course when we humans became

conscious that self-adjusting capability *must* have clashed with our already established dictatorial instincts' orientations that had been managing our lives prior to us becoming conscious.

[111] Many thinkers, in fact many of the greatest thinkers in history, such as Moses with his Garden of Eden story, and Plato in much of his writing, recognised that the emergence of our conscious mind is what led to our departure from our species' original instinctive state of cooperative, selfless and loving innocence—so they were thinking honestly enough to get at least part way to the explanation for our corrupted condition. For example, as mentioned in paragraph 50, Plato wrote of a time when we lived in a **'blessed and spontaneous life...**[where] **neither was there any violence...or quarrel...**[and they had] **no memory of the past** [in other words, we lived in a pre-conscious state]' (*The Statesman*, c.350 BC; tr. B. Jowett, 1871, 271-272). In Video/Freedom Essay 4 (which is further significantly elaborated on in Freedom Essay 53) I present a description of many of these great thinkers from both ancient and contemporary times who have recognised the instinct and intellect *elements* involved in producing the human condition. In addition to Moses and Plato, they include Hesiod from ancient times, and from more recent times, Eugène Marais, Laurens van der Post, Nikolai Berdyaev, Erich Neumann, Paul MacLean, Arthur Koestler, Julian Jaynes, Christopher Booker, Erich Fromm, William Wordsworth, Ralph Waldo Emerson, William Blake, John Milton, Robert A. Johnson and Bruce Chatwin. As was mentioned in paragraph 88, Richard Heinberg's 1990 book *Memories & Visions of Paradise* summarises how universal the acknowledgement of the instinct and intellect *elements* involved in producing the human condition has been: **'Every religion begins with the recognition that human consciousness has been separated from the divine Source, that a former sense of oneness...has been lost...everywhere in religion and myth there is an acknowledgment that we have departed from an original...innocence and can return to it only through the resolution of some profound inner discord... the cause of the Fall is described variously as disobedience, as the eating of a forbidden fruit** [from the tree of knowledge]**, and as spiritual amnesia** [forgetting, blocking out, alienation, denial, psychosis]' (pp.81-82 of 282).

[112] However, while many of the great thinkers in history have recognised the main *elements* of our instinct and intellect involved in producing the human condition, it is the inevitable clash between them that is obviously, now that it has been explained, the reason WHY the presence of those elements led to the psychologically upset state of the human condition. As Heinberg wrote, we could only **'return to'** our **'original'** healthy, upset-free state of **'oneness'** **'through the resolution of some profound inner discord'**, and it is precisely that **'resolution of'** humanity's **'profound inner discord'** that my Instinct vs Intellect explanation has supplied.

[113] Finally, before going on to Part 2, for the benefit of the reader it should be said that since there is a great deal to think about and digest in this massive paradigm-shifting, all-explaining but at the same time all-exposing explanation of the human condition, there is a need for patience and perseverance when reading, watching or listening to presentations of this redeeming and human-race-saving explanation. The reality is that to be able to absorb, understand and appreciate all that is being explained you will very likely need to read this book a number of times, as well as watch/read/listen to *THE Interview* a number of times, and watch/read/listen to the Freedom Essays a number of times, and also re-read or re-listen to the definitive presentation about the human condition and all its manifestations in *FREEDOM*.

[114] Importantly, much more will be said about how we cope with the sudden arrival of the all-redeeming but at the same time all-exposing truth about our corrupted human condition in the coming Parts 2.9 and 2.10.

Part 2

Extended Description of the human condition

Part 2.1 Some of the topics to now be looked at include the 'Deaf Effect', the depth of our anger, and how we cope with exposure of our corrupted human condition

[115] What has been presented so far in this analysis of the human condition is the description of what the human condition actually is, why we have lived in mortal fear of the human condition, why advances in science have made it possible to explain the human condition, and how that explanation frees the human race from the agony and horror of the human condition.

[116] Now that we have the redeeming understanding of the human condition, which is our mind-upset-for-being-condemned neurosis, and soul-repressed-by-our-mind psychosis because of our soul's condemnation of our mind, a whole world of understandings of us humans opens up. When the lights of understanding are turned on in the dark world of denial of our immensely mind-upset and soul-repressed human condition, and of all the human condition-avoiding false 'explanations' for virtually every aspect of human life, there is so much about us humans that we can finally truthfully and compassionately and relievingly understand and make sense of. Some of the more important insights that haven't as yet been described but will be looked at in this 'Extended Description of the human condition' are the difficulty of the 'Deaf Effect' that most people suffer from when they first try to read or listen to analysis of the human condition; why analysis of the human condition can initially cause an extremely angry reaction; and how everyone can cope with the all-redeeming and all-liberating but at the same time all-exposing truth about our immensely corrupted, mind-mad and soul-dead human condition.

[117] Firstly, to go over our basic situation so that we can appreciate how much mental upset and soul-repression we, our conscious thinking self, have been practicing. What happened was that we humans became fully conscious some 2 million years ago, and then when our conscious mind naturally began experimenting in managing our life from a basis of understanding the world, our already established, inflexible, fixed, dictatorial instinctive orientations to the world in effect criticised our conscious mind's experiments because, in undertaking those experiments, our conscious mind was not behaving in accordance with those instinctive orientations. What happened then was that this criticism upset our conscious mind, causing it to become defensively angry towards the criticism, determined to prove it undeserved, and to blocking the criticism out of its thoughts—we, our conscious thinking self, became upset angry, egocentric and alienated. The resulting competitive, selfish and aggressive behaviour drew further condemnation from our particular cooperative, selfless and loving moral instincts, which made our conscious mind even more defensively angry, egocentric and alienated. Our divisive competitive, selfish and aggressive behaviour also made us feel condemned for being out of step with Integrative Meaning or 'God'! We were living in a double and triple whammy situation of rapidly escalating mental upset or neurosis, and soul-repressing psychosis that could only end when our anger, egocentricity and alienation-producing search for knowledge finally led to the finding of the redeeming good reason for why we defied our instincts and became competitive, selfish and aggressive and by so doing corrupted or destroyed our species' original cooperative, selfless and loving instinctive self or soul's world—which is the redeeming Instinct vs Intellect explanation of the human condition that the scientific advances that have been made in understanding how genes and nerves work have finally made it possible to present.

Part 2.2 The two main aspects of life under the duress of the human condition

[118]Importantly, we can see that there were two main aspects to what was happening in our minds.

[119]Firstly, there was the depressing shame and guilt our conscious thinking mind felt for becoming competitive, selfish and aggressive and destroying the innocent cooperative, selfless and loving world of our soul. Secondly, there was a resentful angry feeling that—even though we, our conscious mind, couldn't explain it yet—there was a good reason for why we had become so angry, egocentric and alienated and destroyed the cooperative, selfless and loving world of our soul. We didn't feel 'mad as hell' about our situation for no reason; we didn't attack, try to prove wrong, and block out any criticism we encountered for no reason; we didn't become angry, egocentric and alienated without cause—it was because we felt we were being *unjustly* condemned by the whole damn world!

[120]So our mind felt deeply ashamed and guilty for destroying our cooperative, selfless and loving instinctive self or soul's world, and, at the same time, our mind felt deeply resentful and angry from feeling unjustly condemned as soul-destroying bad, unworthy, guilty, evil and sinful when we intuitively felt we were actually the complete opposite of bad, unworthy, guilty, evil and sinful—namely truly wonderful, good and meaningful; in fact, not just good but a hero of the story of life on Earth! We, our conscious mind, felt immense shame and guilt, but also a deep-seated angry resentment because we believed in our heart of hearts that we weren't actually bad people. So these two polar opposite feelings and their effects have in truth been dominating our lives: extremely depressing shame and raging anger. Basically, our mind didn't know whether we were bad or good: we felt that we were bad, but we also sensed we were good! We couldn't reconcile with understanding which of the bad or good assessments was true; we couldn't reconcile the human condition—which thank goodness we at last now can!

[121]So that has been our situation for some 2 million years, not knowing if we are bad or good—basically, in our subterranean thoughts, we have been worried sick every moment of the day over whether we are evil monsters or sublimely wonderful beings! Even though we have constantly tried to block out that worry by distracting our mind, smiling and laughing and

putting on a brave front, and by finding as much positive reinforcement as we could to quench and hold at bay our frustration and anger; those were the underlying preoccupations in our mind—avoid depression and relieve our anger!

[122] And since this has been going on for some *2 million years*, how much deep subconscious guilt, shame and depression, and deep subconscious volcanic frustration and anger, must now exist within us humans! While we have learnt to hold the subconscious depression at bay with continual efforts to deny it and find ways to delude ourselves that we are happy, buoyant beings, depression must actually be a dominant feature of our lives. And, with regard to our anger, while we have tried to restrain and conceal it—to 'civilise' it as we refer to this restraint—sometimes, when we could no longer find a way to restrain our anger it expressed itself, hence our capacity for shocking acts of cruelty, sadism, sexual depravity, hate, murder and war. So, despite our brave and heroic efforts to stay away from depression and to quell and restrain our anger, under the surface there has existed within us a deep dark shadow of depression, and, at the same time, a boiling-with-anger rage for having to live with what we felt was extremely unjust condemnation. Depression and anger have been the deeper, 'behind the scenes' features of human life—which, now that we have found the redeeming and reconciling understanding of our corrupted condition, can, thank goodness, all subside and disappear from human life forever!

Part 2.3 How depressing the subject of our corrupted human condition has been

[123] With regard to the DEPRESSION, the following are some examples of how excruciatingly depressing it has been for virtually everyone if they allowed their minds to think truthfully about the world's and their own soul-corrupted condition while it wasn't able to be explained and understood—and therefore, as I will describe shortly, why, generation after generation, almost everyone during their early adolescence had no choice but to resign themselves to living in determined denial of the unbearably depressing issue of our species' and of their own soul-destroyed condition.

[124] This is a description of how depressed the French philosopher and scientist René Descartes became when he confronted the horror of his and the human race's corrupted condition: **'So serious are the doubts into which I have been thrown…that I can neither put them out of my mind nor see any way of resolving them. It feels as if I have fallen unexpectedly into a deep whirlpool which tumbles me around so that I can neither stand on the bottom nor swim up to the top'** (*Second Meditation*, 1641; tr. J. Cottingham, 1984).

[125] And this is a member of the public's description of the extreme depression he experienced when he tried to confront the human condition: **'I felt the worst fear I have ever known. Fear doesn't even go close to expressing it. What do you suppose you do when you find the most fearful thing you'll ever encounter is yourself'** (see par. 1185 of *FREEDOM*). Yes, when in a rare moment of perfect clarity a person sees the contrast between how all-loving, all-harmonious and all-sensitive our species once was, which our instinctive soul has the memory of, and how horrifically soul-corrupted, angry, egocentric and alienated they and virtually everyone else now is, the contrast is so great that, without the redeeming explanation for our corrupted condition, its revelation has been **'fearful[ly]'**, unbearably, even suicidally depressing.

Carl Jung (1875–1961)

[126] And this is the deadly accurate description the famous psychoanalyst Carl Jung gave of the **'rare and shattering experience'** that could occur in moments when a person dropped their mental guard and saw their seemingly **'absolute evil'** condition that they couldn't explain and understand: **'When it** [the **shadow** of our corrupted condition] **appears…it is quite within the bounds of possibility for a man to recognize the relative evil of his nature, but it**

is a rare and shattering experience for him to gaze into the face of absolute evil' (*Aion: Researches into the Phenomenology of the Self*, 1959; tr. R. Hull, *The Collected Works of C.G. Jung*, Vol. 9/2, p.10).

[127] And the renowned German polymath Johann Wolfgang von Goethe's comment about the aphorism **'Know thyself? If I knew myself, I'd run away'** (*Elective Affinities,* 1809) is entirely understandable while we couldn't understand ourself, explain the human condition, know why we corrupted our soul, know that we were actually good and not bad!

[128] The philosopher Søren Kierkegaard's **'analysis on the nature of despair is one of the best accounts on the subject'** (Wikipedia; see www.wtmsources.com/137) — with 'the nature of despair' being as close as the reviewer could go in referring to the worse-than-death, suicidal depression that the subject of the human condition has historically caused humans, but which Kierkegaard managed to give such an honest account of in his aptly titled 1849 book, *The Sickness Unto Death*: **'the torment of despair is precisely the inability to die** [and end the torture of our previously unexplained human condition]**…that despair is the sickness unto death, this tormenting contradiction** [of our 'good and evil', human condition-stricken lives], **this sickness in the self; eternally to die, to die and yet not to die'** (tr. A. Hannay, 1989, p.48 of 179).

[129] So when the great English poet Gerard Manley Hopkins wrote about the unbearably depressing subject of the human condition in his (also) aptly titled 1885 poem *No Worst, There Is None*, his words, **'O the mind, mind has mountains; cliffs of fall, frightful, sheer, no-man-fathomed'**, did not exaggerate the depth of depression almost everyone faced if they allowed their mind to think about the human condition while it was still to be 'fathomed'/understood/answered.

[130] As mentioned earlier in paragraphs 28-29, in 1988 *TIME* magazine published a deeply reflective article about Alan Paton's favourite pieces of literature in which Paton said, **'I would like to have written one of the greatest poems in the English language — William Blake's "Tiger, Tiger Burning Bright", with that verse that asks in the simplest words the question which has troubled the mind of man — both believing and non believing man — for centuries: "When the stars threw down their spears / And watered heaven with their tears / Did he smile his work to see? / Did he who made the lamb make thee?"'** (*TIME*, 25 Apr. 1988). As I pointed out, Blake's poem (from his book *Songs Of Innocence and Of*

Experience) raises that fundamental question involved in being human of how could the mean, cruel, indifferent, selfish and aggressive 'dark side' of human nature—represented by the **'Tiger'**—be both reconcilable with and derivative of the same force that created **'the lamb'** in all its innocence?

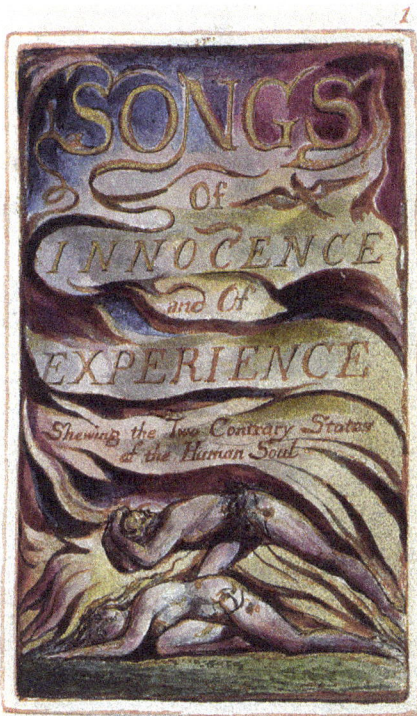

The cover of William Blake's 1794 *Songs Of Innocence and Of Experience: Shewing the Two Contrary States of the Human Soul*

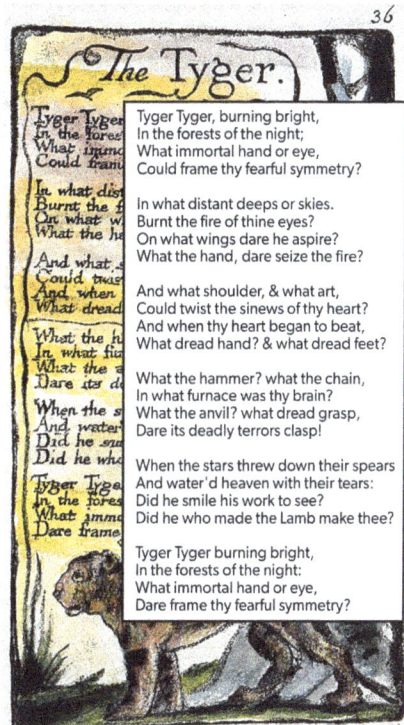

The Tyger by William Blake, from *Songs Of Innocence and Of Experience*

[131]Having now explained the underlying depression in human life, it is appropriate to refer to Blake's extraordinarily truthful poem again, in particular to the poem's opening lines, **'Tyger Tyger, burning bright, in the forests of the night'**, because we can now understand that they refer to humans' great fear and resulting determined denial of the unbearably depressing issue of our seemingly highly imperfect, 'fallen' or corrupted, soul-destroyed state or condition—a subject we have consciously repressed and yet one that has been **'burning bright, in the forests of the night'** of our

deepest awareness. As Blake so honestly described in his poem the horrific depression that the issue of the human condition has until now caused: **'what' 'eye' 'could'** be expected to look at the **'fearful'** subject, **'what' 'hand'** would **'dare seize the fire'** that **'could twist the sinews of thy heart?'**; the terrible **'hammer'**, the **'furnace'** in **'thy brain'**, no one can possibly **'dare its deadly terrors clasp!'** Again, the very heart of the unbearably depressing issue lies in the line, **'Did he who made the lamb make thee?'** — a rhetorical question disturbing in its insinuation that we monstrously mentally upset or neurotic, and horrifically soul or psyche-repressed or psychotic humans are wholly unrelated to **'the lamb'** in all its innocence!

[132] As Paton truthfully identified, the great, fundamental, underlying question that **'has troubled the mind of man'** has always been, are humans part of God's **'work'**, part of **'his'** purpose and design, or aren't we? As was explained earlier in paragraph 60, the Integrative Meaning of life is the Negative Entropy-driven development of order of matter where selfless consideration for the maintenance of the larger wholes is needed to hold the larger wholes together — which, as I pointed out is a critically important truth that hasn't been able to be admitted while we couldn't truthfully explain our *divisively* behaved, competitive, selfish and aggressive human condition. In fact, as I pointed out, Integrative Meaning has been so unbearably condemning of us divisively behaved humans that we made it part of a realm unrelated to our Earthly existence by deifying it as a 'God'. So yes, underneath our brave facades has lurked the depression of being unable to answer the fundamental questions of why haven't we lived in accordance with Integrative Meaning, why haven't we been part of God's **'work'**, *are we good or are we evil?*

Part 2.4 How angry the human condition has made us

[133] With regard to the depth of ANGER in us humans, the following is a cartoon by the brilliantly honest and insightful Australian cartoonist Michael Leunig that accurately recognises humans' deep anger from feeling unjustly condemned for destroying our species' original innocent instinctive self or soul's idyllic cooperative, selfless and loving, 'Garden of Eden' home.

Cartoon by Michael Leunig that appeared in Melbourne's *The Age* newspaper on 31 Dec. 1988

Drawings by Jeremy Griffith, with deeply appreciative deference to Michael Leunig, Jul. 2009

[134] Firstly, as was explained earlier in paragraph 85, the Biblical story of Adam and Eve, which is truthfully titled in the Bible **'The** [soul-corrupting] **Fall of Man'**, describes how Adam and Eve (we humans) lived originally in a **'Garden of Eden'** paradise of peaceful, loving innocence, and then

we took the **'fruit' 'from the tree of...knowledge'** and were **'disobedient'**. In other words, we developed a conscious mind and free will. But in that *pre-scientific* story it says Adam and Eve then became **'evil'** perpetrators of **'sin'** because, as we now understand, they became angry, egocentric and alienated, and as a result were **'banished...from the Garden of Eden'** state of cooperative, selfless and loving innocence, an expulsion that, with the benefit of science, we can explain and understand was not justified and so can end!

[135] While we have only now been able to explain the good reason for our corrupted condition, we humans have, as I have been pointing out, always intuitively believed that we weren't **'evil'** for becoming corrupted and didn't deserve to be **'banished'**, and <u>Leunig has perfectly captured the frustrated, angry effects of feeling so unjustly treated</u>. As I explain about this amazingly revealing cartoon by Leunig in paragraphs 288-290 of *FREEDOM*, we see in the cartoon the <u>psychotic</u> (soul-destroying) and <u>neurotic</u> (mind-furious) anger that the feeling of being unjustly condemned has caused us—vengeful Adam chain-sawing and razing the Garden. The guardian angel is in tears at the wanton destruction, and we can see that Eve is similarly distressed by his actions. (<u>This aforementioned (in paragraph 54) lack of empathy by women for men's immensely upsetting and astronomically heroic battle to defy the ignorant condemnation from our instinctive self, which Leunig has so honestly portrayed here, is explained in chapter 8:11B of *FREEDOM*—and I should say again that the extremely important revelation in that chapter of the different roles that men and women have had in humanity's unbreakably brave struggle to solve the human condition is yet another insight that wasn't possible while we couldn't explain the human condition</u>.) But Adam's expression and body language shows the enormous relief and satisfaction his retaliation brings him. In giving the guardian angel 'the finger' in the eighth frame, Adam is symbolically saying, *'Go to hell you bastard for unjustly condemning me!'*

[136] But, above all, in the expression of extreme anger on Adam's face, Leunig has revealed just what 2 million years of being unjustly condemned by the *whole* world has done to us humans. Yes, again, since the sun, the rain, the trees and the innocent animals are all friends of our original innocent, instinctive self or soul, through that association they too have condemned us, which is why Adam's (all humans', but especially men's) retaliation against nature for its unjust condemnation of him (of men

especially) has left the whole natural world such a wasteland! We (men especially) burnt the scrub, tore down the trees and dumped rubbish, pollutants and cement over what was left of nature, and, as explained in paragraphs 778-782 of *FREEDOM*, we murdered animals because their innocence condemned us! For example, the diary of the legendary 'white hunter', the suitably named J.A. Hunter, reveals that he dispatched '**966 Rhinos**' from '**August 29th 1944 to October 31st 1946**' (Peter Beard, *The End of the Game*, 1963, p.137 of 280). That's the equivalent of nearly 10 rhinoceroses every week for more than 2 years that he shot to death! Similarly, there has been '**5,843 Sq km of Amazon rain forest reportedly lost to deforestation from August 2012 to July 2013; activists blame the 28% rise in one year on looser environmental laws**' (*TIME*, 2 Dec. 2013)—the deeper truth being that if we humans weren't so hateful and couldn't-care-less we wouldn't destroy pristine rainforest like this.

Photo of a shooter posing with a giraffe he has just shot to death. Note the satisfaction on his face, and the crumpled dead victim of his get-even-with-innocence-for-its-unjust-condemnation, triumphant brutalisation!

[137]Incidentally, we can see here how, if we really wanted to save the environment, that 'hugging trees' and 'patting dolphins' wasn't going to do it. Pretending to be loving and kind and considerate could make us feel good but it was never going to actually fix anything. To 'save the planet' we had to find the *understanding* that would end the underlying upset in us humans. We had to confront the issue of *us*, *the human condition*,

not find ways to delude ourselves we were good so we didn't have to confront that issue. As I explain in my book *Death by Dogma*, pretence, delusion and escapism got us nowhere; in fact, it made the situation the human race is in much, much worse because it hid the real issue of the immensely mentally upset or <u>neurotic</u> and immensely soul-repressed or <u>psychotic</u> state of the human condition. So yes, while we have learnt to conceal how neurotic and psychotic we are—learnt 'civility'—underneath that facade of restraint and delusion lies the level of anger Leunig has portrayed.

[138] It was with great reverence to Leunig that I took the liberty of drawing three more frames in his Garden of Eden cartoon to complete the story by showing how having finally found the redeeming understanding of ourselves we are able to metaphorically call the guardian angel back and show it that redeeming insight, which causes the guardian angel to cry in sympathy and take us back to the Garden of Eden! <u>What a wonderful metaphorical depiction the whole cartoon now is of our anger for being unjustly condemned for corrupting our original cooperative, selfless and loving, innocent instinctive self or soul, and how we can now return to that fabulous peaceful and loving-of-all-things way of living!</u>

[139] So to again consider Leunig's depiction of Adam (males), whose expression and body language convey the enormous relief and satisfaction he feels from retaliating against his feeling of being unjustly treated. Being trapped in such a volcanic situation, it really is no wonder that the whole human race, but men in particular, have led such an evasive, escape-any-self-confrontation, denial-practising, lying, avoid-any-criticism, egocentric, soul-denying-and-repressing, pretend-everything-is-fine-when-it-actually-isn't, neurotic and psychotic, alienated-to-the-core, superficial and artificial, self-centred, indifferent-to-other-people's-situations, greedy, power, fame, fortune and glory-seeking existence! <u>As I said in *THE Interview*, we humans have *had to* smother ourselves with *material* glory while we lacked the *spiritual* glory of being able to compassionately understand ourselves.</u>

Material reinforcement had to sustain us until we found spiritual reinforcement, understanding

[140] Self-distracting, escapist materialism and the capitalism that supplies it, have been the necessary poor substitutes for spiritualism, the ability of our conscious mind to understand why it corrupted the magical, all-harmonious, all-loving and all-sensitive world of our soul! Consider what we had lost—as Sir Laurens van der Post wrote about the sensitivity of the relatively innocent Bushmen or San people of the Kalahari desert: **'He and his needs were committed to the nature of Africa and the swing of its wide seasons as a fish to the sea. He and they all participated so deeply of one another's being that the experience could almost be called mystical. For instance, he seemed to *know* what it actually felt like to be an elephant, a lion, an antelope, a steenbuck, a lizard, a striped mouse, mantis, baobab tree'** (*The Lost World of the Kalahari*, 1958, p.21 of 253). And how wonderfully revealing of our lost sensitivity was the quote from Fyodor Dostoevsky from his beautifully titled book *The Dream of a Ridiculous Man* [that one day we might resurrect our innocent, all-loving and all-sensitive self] that I presented in *THE Interview*!

[141] And of course we haven't only taken out our anger on nature, we have also taken it out on each other, with endless wars, murders and atrocities committed against our fellow humans.

Humans' inhumanity to humans!

Part 2.5 The extent of our mind-mad and soul-dead, human-condition-stricken existence

[142] Overall, our conscious mind's insecurity about our corrupted condition has been so great that our mind has been deeply preoccupied all the time with finding relief from the horror of that predicament! As I said, pretending we love animals or love a beautiful sunset or painting or whatever, makes us feel good, but in truth our mind has been preoccupied elsewhere, owned by a terrible dead weight of depression and a smouldering fury of anger! Thank goodness a million, million times

over then that science has enabled the redeeming Instinct vs Intellect explanation of our corrupted human condition to be assembled, and our insecure angry, egocentric and alienated, mind-mad and soul-dead way of living to come to an end!!

Cartoon by Michael Leunig, Melbourne's *The Age* newspaper, 8 Oct. 1988

[143] This cartoon is another brilliantly honest and revealing masterpiece by Michael Leunig. Here he truthfully depicts all the horrors of the human condition. It is not a picture of a lovely ordered city park where people peacefully and happily enjoy themselves, as we human-condition-avoiding people delude ourselves the world we have created is like. Rather, it shows a mother and child approaching the **'Gardens of the Human Condition'** with an expression of bewildered dread on the face of the mother, and in the case of the child, wide-eyed shock. Yes, as Leunig cleverly intimates, our world is no longer an innocent Garden of Eden, but a devastated realm of human-condition-stricken, mind-mad and soul-dead humans where 'in̲humanity' reigns. With this brilliantly honest masterpiece,

Leunig has boldly revealed the truth that we humans are a brutally angry, hateful, destructive, arrogant, egocentric, selfish, mad, lonely, unhappy and depressed species. He has people fighting, beating and strangling each other, drunk out of their minds, depressed, lonely, crying, hiding and suiciding, going mad, and egocentrically holding forth—reflecting every aspect of the human condition.

[144] Yes, as the main character in the 2005 film *The White Countess* noted, '**What we see out there** [in the world] **is chaos; mistrust, deception, hatred, viciousness—chaos—there's no broader canvas out there, nothing that man can go and compose a pretty picture on.**' The polymath Blaise Pascal was even more damning in his depiction of the human condition, when, in 1669, he spelled out the full horror of our contradictory nature, writing, '**What a chimera then is man! What a novelty, what a monster, what a chaos, what a contradiction, what a prodigy! Judge of all things, imbecile worm of the earth, repository of truth, a sewer of uncertainty and error, the glory and the scum of the universe!**' (*Pensées*, 1669). William Shakespeare was equally revealing of the paradoxical true nature of the human condition when in 1603 he wrote, '**What a piece of work is a man! How noble in reason! How infinite in faculty!... In action how like an angel! In apprehension how like a god! The beauty of the world! The paragon of animals! And yet, to me, what is this quintessence of dust?** [Brutal and barbaric] **Man delights not me**' (*Hamlet*, 1603)!!

[145] So the 17 goals of '**The** [United Nations'] **Global Goals' 'Movement'** that each '**global citizen**' and '**every school**' is currently being told to '**tell everyone about**' and '**make famous**' (namely to address '**poverty, hunger, well-being, education, gender equality, clean water, clean energy, decent work, infrastructure, inequality, sustainability, responsible consumption, climate action, life in water, life on land, peace and justice, partnerships for the goals**') *hugely* trivialise our species' plight—because all these goals focus *only* on the symptoms of the human condition; they embody the 'first interpretation' of the human condition that was introduced at the beginning of this book, an interpretation that we now know is completely superficial. To stop the destruction of our world and the disintegration of society that is happening everywhere we look we have to fix the *cause* of the problems at its source, which is *us humans, our neurotic and psychotic existence*. WE are the problem; our out-of-control egocentric, selfish, competitive and ferociously vicious, mean and aggressive behaviour. The cartoonist Walt Kelly spoke the truth when he had Pogo, his comic strip hero, say, '**We have met the enemy and he is us**' (1971).

[146] Yes, the underlying, REAL question that had to be answered if we were ever to find the mentally redeeming and soul-relieving understanding of ourselves was why are we humans the most brilliantly clever of creatures — **'god'-'like'** in our **'infinite' 'faculty'** of **'reason'** and **'apprehension'**, a **'glor**[ious]**'**, **'angel'-'like' 'prodigy'** capable of being a **'judge of all things'** and a **'repository of truth'** — also the meanest, most vicious, most capable of inflicting pain, cruelty, suffering and degradation? Why are humans so choked full of volcanic frustration, anger and hatred — the species that behaves so appallingly that we seem to be **'monster**[s]**'**, **'imbecile**[s]**'**, **'a sewer of uncertainty and error'** and **'chaos'**, the **'essence'** of **'dust'**, **'the scum of the universe'**? *That* is what the issue of the human condition *really* is — and what beyond-description-relief it is that we can finally understand and end the terror and horror of the human condition! As Professor Stuart Hurlbert, Emeritus Professor of biology at San Diego State University, has said, **'With the world in chaos what's really needed is someone to finally make sense of humans' seemingly mad behaviour — which is actually what Australian biologist Jeremy Griffith, who I consider to be the newest Charles Darwin, has done. His world-saving breakthrough presents the redeeming and psychologically healing understanding of ourselves, which is the fundamental paradigm shift we need to save the human race.'**

'How infinite in faculty'—at 830m high, the Burj Khalifa in Dubai is the world's tallest skyscraper

But **'what a monster'**—war in Kyiv, Ukraine, 2022

[147] Truly, with every day bringing with it more alarming evidence of the turmoil of the human situation, the issue of the human condition is the *ONLY* question confronting the human race, because its solution has become a matter of critical urgency. Conflict between individuals, 'races', cultures, religions and countries abounds (and by 'races' I mean groups of people whose members have mostly been together a long time and are thus relatively closely related genetically—people who have a shared history). There is genocide, terrorism, extreme greed and dysfunction, crisis-level migration of peoples, starvation, runaway diseases, environmental devastation, gross inequality, 'racial' and gender oppression, polarised politics, rampant corruption and other crimes, drug abuse, family breakdown, and epidemic levels of obesity, chronic anxiety, depression, unhappiness and loneliness—all of which are being rapidly exacerbated by the exploding world population, and a resulting exponential rise everywhere in the anger, egocentricity and alienation of our mind-upset, neurotic and soul-repressed, psychotic human condition. Improved forms of management, such as better

laws, better politics and better economics—and better self-management, such as new ways of disciplining, suppressing, organising, motivating or even transcending our extremely troubled natures—have all failed to end the march towards ever greater levels of alienation, devastation and unhappiness. In short, the situation is now *so* grim the human race *IS*, in fact, entering *end play* or *end game*, where the Earth cannot absorb any further devastation from the effects of our mad behaviour, nor the human body cope with any more debilitating stress, or, most particularly, our mind endure any more agony, any more alienated neurosis and psychosis. The journalist Richard Neville got the truth up when he wrote that '**the world is hurtling to catastrophe: from nuclear horrors, a wrecked eco-system, 20 million dead each year from malnutrition, 600 million chronically hungry...All these crises are man made, their causes are psychological. The cures must come from this same source; which means the planet needs psychological maturity fast. We are locked in a race between self destruction and self discovery**' (*Good Weekend*, *The Sydney Morning Herald*, 14 Oct. 1986; see www.wtmsources.com/167). Yes, our species has come to the critical juncture where *ONLY* '**self discovery**'—reconciling, ameliorating, '**psychological**[ly]' healing understanding of ourselves—could save us from '**self destruction**', and thank goodness that is what the human race has with the redeeming Instinct vs Intellect understanding of our mind-mad and soul-dead human condition!

Part 2.6 Our denial of the human condition and its alienating effects

[148] Since the truth of our species' 2-million-year innocence-destroyed, soul-corrupted, 'fallen' human condition has been such an absolutely unbearably depressing truth, the only way we have been able to cope with it was to live in almost complete denial of the truth that our species ever did once live in a cooperative, selfless and loving innocent state, and in almost complete denial that we suffer from any form of guilt-ridden, neurotic and psychotic condition! Along with extreme denial, extreme pretence and extreme delusion have been our stocks in trade—they are what have got us through each day.

Drawing by J. Griffith © 1991 Fedmex Pty Ltd

Denial protected us from unbearably depressing confrontation with the
truth of our innocence-destroyed, soul-corrupted, 'fallen', massively
angry, egocentric and alienatedhuman condition (the crown represents
the egocentric, 'I am a legend' delusion we adopted)

[149] The preeminent Scottish psychiatrist R.D. Laing extraordinarily
bravely and truthfully recognised how fearful we have been of the
unbearably depressing subject of our soul-corrupted condition while
we couldn't explain it, and therefore how committed we have been to
blocking that truth out by denying our species ever lived in an original
cooperative, selfless and loving innocent state, and, as a result of all
this mental blocking out and soul-repression, how estranged from or
separated from or alienated from our species' original cooperative,
selfless and loving instinctive self or soul we are. In the 1960s he wrote
these absolutely astonishingly honest words that basically reiterate
everything that is said in this presentation I am giving about what the
human condition really is: 'The requirement of the present, the failure of the
past, is the same: to provide a thoroughly self-conscious and self-critical human
account of man [p.11 of 156] ...Our alienation goes to the roots. The realization
of this [truth] is the essential springboard for any serious [truthful, effective,
penetrating] reflection on any aspect of present inter-human life...We are all
murderers and prostitutes...We are bemused and crazed creatures, strangers
to our true selves, to one another...We are born into a world where alienation
awaits us. We are potentially men, but are in an alienated state [pp.11-12] ...the

ordinary person is a shrivelled, desiccated fragment of what a person can be. As adults, we have forgotten most of our childhood, not only its contents but its flavour; as men of the world, we hardly know of the existence of the inner world [p.22] ...Our capacity to think, except in the service of what we are dangerously deluded in supposing is our self-interest, and in conformity with common sense, is pitifully limited: our capacity even to see, hear, touch, taste and smell is so shrouded in veils of mystification that an intensive discipline of un-learning is necessary of *anyone* before one can begin to experience the world afresh, with innocence, truth and love [p.23] ...The condition of alienation, of being asleep, of being unconscious, of being out of one's mind, is the condition of the normal man. Society highly values its normal man [expects everyone to become resigned to living in denial of the human condition]. It educates children to lose themselves [become resigned] and to become absurd, and thus to be normal [p.24] ...Each child is a new beginning, a potential prophet [denial-free, truthful, effective thinker], a new spiritual prince, a new spark of light, precipitated into the outer darkness [that we are living in] [p.26] ...Torn, body, mind and spirit, by inner contradictions, pulled in different directions, Man cut off from his own mind, cut off equally from his own body – a half-crazed creature in a mad world...We are all implicated in this state of affairs of alienation [pp.46-47] ...we are driving our children mad more effectively than we are genuinely educating them [p.87] ...*If the formation is itself off course, the man who is really to get 'on course' must leave the formation* [p.99] ...Schizophrenia used to be a new name for dementia... that was supposed to overtake young people in particular [see the description of Resignation coming up in paragraph 160], and to be liable to go on to a terminal dementia. Perhaps we can still retain the now old name, and read into it its etymological meaning: *Schiz* – "broken"; *Phrenos* – "soul or heart" [p.107] ...we have all reason to be insecure. When the ultimate basis of our world is in question, we run to different holes in the ground, we scurry into roles, statuses, identities, interpersonal relations [p.108] ...The outer divorced from any illumination from the inner is in a state of darkness. We are in an age of darkness. The state of outer darkness is a state of sin – i.e. alienation or estrangement from the inner light [p.116] ...between *us* and It [our true selves or soul] there is a veil which is more like fifty feet of solid concrete. *Deus absconditus.* Or we have absconded... To adapt to this world the child abdicates its ecstasy...There is a prophecy in [the Old Testament book of] Amos that there will be a time when there will be a famine in the land, "not a famine for bread, nor a thirst for water, but of *hearing* the words of the Lord [words of truthfulness]." **That time has now come to pass. It is the present age** [p.118] **…How do you plug a void plugging a void? How to inject nothing into fuck-all? How to come into a gone world?…I do assure you. The dreadful has already happened** [p.153]' (*The Politics of Experience* and *The Bird of Paradise*, 1967). **'We are dead, but think we are alive. We are asleep, but think we are awake. We are dreaming, but take our dreams to be reality.** We are the halt, lame, blind, deaf, the sick. **But we are doubly unconscious. We are** *so* **ill that we no longer feel ill, as in many terminal illnesses. We are mad, but have no insight** [into the fact of our madness]' (*Self and Others*, 1961, p.38 of 192). **'We respect the voyager, the explorer, the climber, the space man. It makes far more sense to me as a valid project – indeed, as a desperately urgently required project for our time – to explore the inner space and time of consciousness** [which we haven't done effectively because of our conscious mind's fear of the human condition]. **Perhaps this is one of the few things that still make sense in our historical context. We are so out of touch with this realm** [so in denial of the issue of our alienated, mind-mad, soul-dead human condition] **that many people can now argue seriously that it does not exist. It is very small wonder that it is perilous indeed to explore such a lost realm** [p.105] **…True sanity entails in one way or another the dissolution of the normal ego, that false self competently adjusted to our alienated social reality: the emergence of the "inner" archetypal mediators of divine power, and through this death a** [transformed] **rebirth, and the eventual re-establishment of a new kind of ego-functioning, the ego now being the servant of the divine, no longer its betrayer** [we can now explain why the ego or intellect *had* to '**betray**' our '**divine**' instinctive self or soul] [p.119] **… the direction we have to take is** *back* **and** *in* [p.137]' (*The Politics of Experience* and *The Bird of Paradise*).

[150] Holy dooley, what an absolutely incredibly brave and honest description of the human condition in a time when there was still no redeeming understanding of our corrupted condition like there is now! And to think I have brought compassionate, relieving and healing explanation and understanding to all the problems that R.D. Laing has so honestly described. What redemption and relief for the human race! And to do that, all I did was think truthfully—as R.D. Laing said, undertake the **'desperately urgently required project for our time – to explore the inner space and time of consciousness',** with '**consciousness**' being a traditional code word for our conscious-mind-troubled human condition.

Peter Davis 1967

The irreverent, disrespectful-of-the-intellectual-world-of-denial,
fearlessly-truth-revealing R.D. Laing (right) in 1967, addressing
the 'Dialectics of Liberation' (whatever that means!) conference

[151] A rare example of someone who managed to penetrate and honestly
depict this **'fifty feet of solid concrete'** wall of alienating denial of the truth
of our tortured human condition was the great Irish artist Francis Bacon.
While people living in denial of what the human condition actually is,
maintaining **'that it does not exist'** as Laing said, typically find Bacon's
work **'enigmatic'** and **'obscene'** (*The Sydney Morning Herald*, 29 Apr. 1992), there is
really no mistaking the agony of the human condition in Bacon's death-
mask-like, twisted, smudged, distorted, trodden-on—alienated—faces,
and tortured, contorted, stomach-knotted, arms-pinned, psychologically
strangled and imprisoned bodies; consider, for instance, his *Study for
self-portrait* (opposite)! Indeed, Bacon actually said that through his
art he sought to puncture the **'veil'** Laing wrote about above, saying,
**'We nearly always live through screens – a screened existence. And sometimes
I think, when people say my work looks violent that perhaps I have from time
to time been able to clear away one or two of the veils or screens'** (David Sylvester,
Interviews with Francis Bacon, 1993, p.82). It is some recognition of the incredible

integrity/honesty/truth in Bacon's work that in 2013 one of his triptychs sold for an astonishing $US142.4 million, becoming (at the time) **'the most expensive work of art ever sold at auction, breaking the previous record, set in May 2012, when a version of** [the Norwegian painter] **Edvard Munch's** *The Scream* [another exceptionally honest, horror-of-the-human-condition-revealing painting that I will be referring to again shortly in paragraph 170] **sold for $119.9 million'** (*TIME*, 25 Nov. 2013).

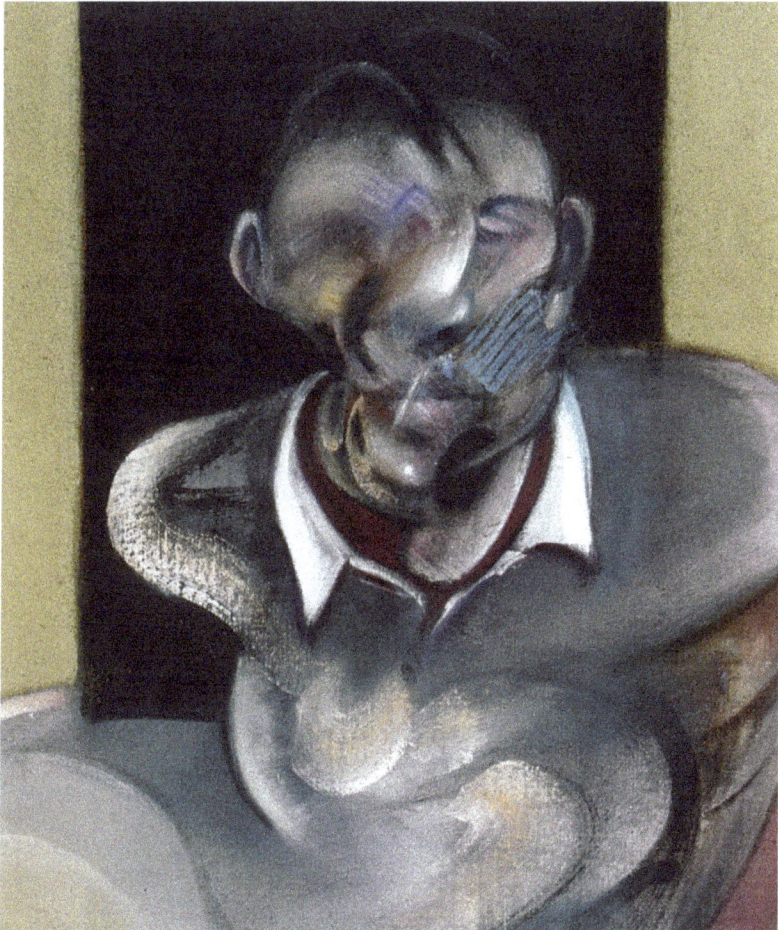

Detail from Francis Bacon's *Study for self-portrait*, 1976

[152]Another example of an extremely honest description like R.D. Laing's of our soul-corrupted, mad-with-pain-and-anger human condition was given by Jesus Christ—one of the most honest, denial-free, sound minds in recorded history—when he said, '**O bitterness of the fire** [Christ is using the metaphor of fire to describe the extremely psychologically upset anger, egocentricity and alienation in humans] **that blazes in the bodies of men and in their marrow, kindling in them night and day, and burning the limbs of men and making their minds become drunk and their souls become deranged... Woe to you, captives, for you are bound in caverns** [in the next Part 2.7 we will see how Plato also described humans as being captive prisoners in a dark cave]! **You laugh! In mad laughter you rejoice** [pretend you're happy and all is well when the truth is you're full of suffering and despair]! **You neither realize your perdition, nor do you reflect on your circumstances, nor have you understood that you dwell in darkness and death** [you live in determined denial of your corrupted human condition]! **On the contrary, you are drunk with the fire and full of bitterness. Your mind is deranged** on account of the **burning that is in you, and sweet to you are the poison and the blows of your enemies! And the darkness rose for you like the light, for you surrendered your freedom for servitude** [to denial]! **You darkened your hearts and surrendered your thoughts to folly, and you filled your thoughts with the smoke of the fire that is in you! And your light has hidden in the cloud** [of darkness] **and the garment that is put upon you, you** [pursued] [with deceit]. **And you were seized by the hope that does not exist. And whom is it you have believed? Do you not know that you all dwell among those who** [lie] [... **and you boast**] **as though** [**you had hope**] [that you can achieve satisfaction through winning power, fame, fortune and glory]. **You baptized your souls in the water of darkness! You walked by your own whims!**' This amazingly honest description of the corrupted state of the human condition is from the *Book of Thomas*, a recording of a conversation Christ had with Thomas, one of his disciples. I have written about Christ's phenomenally sound and brave, human-race-saving, lies-defying, truth-defending life in Freedom Essay 39—which I encourage you to read, as many people have, for example, commented that '**This is the best description of Christ I have ever read!**' (Jivka Foster on Facebook, see www.wtmsources.com/215).

Jeremy Griffith's 2004 drawing of Jesus Christ as the **'lamb of God'** (John 1:29, 36),
someone exceptionally innocent, free, centred, secure and natural.

Part 2.7 The problem of the 'Deaf Effect' and the anger that analysis of the human condition initially causes

[153]What now needs to be pointed out—and this follows from what has been said about humans' historic immense fear of the unbearably depressing subject of the corrupted state of our human condition—is that when the explanation of our corrupted human condition is finally found and presented, as it now has been, even though the explanation is redeeming and liberating, trying to read about or listen to discussion of that corrupted condition is initially going to trigger extreme subconscious fear in virtually everyone. They will fear they are being taken back into what they remember as the death zone, the **'deadly terror'** that William Blake described confronting the human condition as causing (see paragraph

131). And this fear will express itself by their mind not tolerating what is being explained. The blocks they have been employing to protect themselves from confronting the human condition will come into play and their mind won't allow what is being talked about to be taken in and absorbed. Then, not aware that their mind is blocking out what is being discussed (we can't block something out and know we're blocking it out or we wouldn't be blocking it out), they will blame the quality of the presentation for the difficulty they are having when reading or listening to it. And that is what happens: people typically say that what is being presented in my books and other presentations is **'impenetrably dense'**, **'confusingly worded and long-winded'**, **'unnecessarily repetitive'**, **'desperately needs editing'**, and that it is **'exceedingly boring and tedious'**, and even that it is **'completely lacking in any substance or meaning'**. Quite often people become so bewildered they request **'an executive summary so I have some idea of what it is that you're trying to say'**! This is well-known to us in the WTM as the 'Deaf Effect' that a great many people experience; at least initially, because with perseverance reading and listening to presentations about the human condition, the redeeming nature of the explanation of the human condition gradually reassures a person's mind that it is finally safe to look into the human condition—and from there they become extremely excited to be able to understand the human condition and from there every aspect of human existence! I need to emphasise that the Deaf Effect is such a serious problem for our project of getting this information out to the world in order to save the human race, it is really THE problem facing the world! For example, one person said, **'Initially when I talked to my family and friends about this the reception was such that I may as well have been talking to the dog!'** Another person similarly said when he tried to share our understandings with others that **'It saddens and frustrates me to say this but such is the wall of denial I have encountered when trying to tell people about Jeremy's incredibly important explanation of the human condition that it has become very clear to me that I may as well be speaking Swahili!'** Yet another person said, **'When I first read this material all I saw were a lot of black marks on white paper.'** Another reader said, **'This stuff is so head on it can be crippling, which, initially at least, can make it hard to get behind what's being said and access the profundity of where it's coming from.'** In the case of a Swiss subscriber who runs Shamanic Healing workshops, he said: **'My healing work isn't possible without Jeremy's biological explanation of the human condition because no matter how much you connect people with the soul, the** mind still questions and needs to understand *why* we carry so much guilt, and Jeremy's work is the medicine for the mind to make the connection between mind and soul. The problem with Jeremy's work is it's so hard to reach people with it who are living in denial of the human condition. The root problem is this guilt that is buried so deep [this is Laing's **'fifty feet of solid concrete'** that humans have resigned themselves to having to adopt in order to protect them from confrontation with the truth of their and the world's corrupted condition]. When we give Jeremy's book to them they put it aside and don't read it because it is too confronting. I mean you face your inner devils and demons on this path and I see many people when the question pops up of are they good or evil, they shy away from the process and try everything to avoid going to that point...so I break down their defences to get them to connect to their inner child using shaman techniques and special herbs...and then give them Jeremy's explanations when they're open'. Possibly, alienation-loosening drugs such as microdoses of psilocybin—which are emerging as promising tools in psychotherapy—will be found to be useful for some people in overcoming the Deaf Effect, but we in the WTM do not advocate such use of drugs because, for one thing, they would require a great deal of testing, monitoring, regulatory approval, etc, before any use could seriously be considered. (See Deaf Effect Management News at www.humancondition.com/deaf-effect-management-news.) I have also heard of another group who are promoting the WTM's explanation of the human condition specifically tell people to '**Just keep reading and listening to Griffith's presentations over and over again until you get it'**, and, as described above, this advice is in line with what we have learnt enables people to overcome the Deaf Effect. For example, the following comment about the effect of persevering is from a post on the WTM's Facebook Group: **'Hello, I'm a first time poster. Something weird is happening – I thought I'd share. I first came across this book [*FREEDOM*] in 2017 when I downloaded the free version. I couldn't get past the first few pages. Then in 2020, I bought the paperback version for myself and my Dad, but still couldn't get through the first couple of pages. It just didn't make any sense, I felt that the sentences were too long and rambling. Reading it made me tired, my mind would wander away every time, so I put it down again. Recently** [in 2023] **after seeing lots of WTM content pop up on social media I started playing the main pinned interviews** [in the Facebook Group] **in the morning as I was getting ready for work. A few tiny bits of information clicked into place, nothing mind blowing, but definitely a few "ah ha" moments. Today, I've been sitting at the bay waiting for my husband to finish a charity bike ride. I bought a coffee** and thought it would be a good location to try reading the book again. In just over an hour, I've gone through 50 pages. I don't even remember why I found it so difficult, all of a sudden it's making sense and I'm enjoying it. It's like I'm reading a different book!' Yes, perseverance works, as this couple found: 'In *FREEDOM* I found the enlightenment I had long dreamed of…[but] was surprised others had difficulty hearing this information…[My wife] Cindy suffered from the "deaf effect". She told me the book was poorly written, repetitive and literally incomprehensible…since I seemed so convinced that this was really the information to save the world, Cindy agreed to keep at it. After many weeks of reading the information together, and openly and honestly discussing it, to both our amazement, Cindy began to not only hear the information but gain an understanding at an even deeper level than what I could initially access.' You can read/watch Bill and Cindy McCaugherty's story at www.wtmcanadawestcoast.com. And you can learn more about the extremely serious problem of the Deaf Effect in Video/Freedom Essay 11.

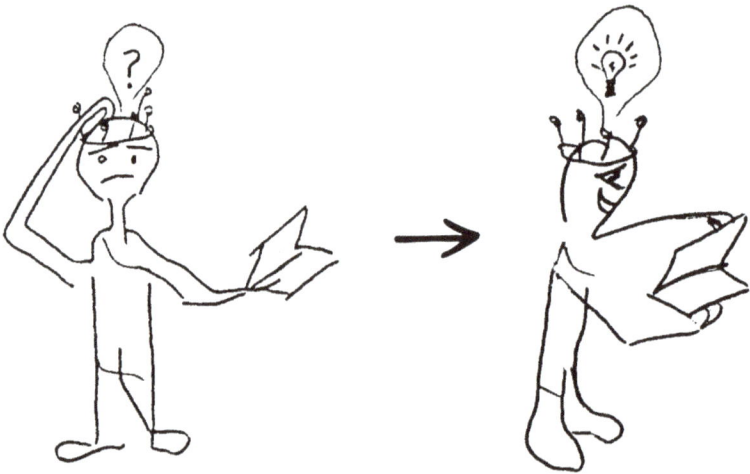

The initial Deaf Effect reaction to description of the human condition
and how that can be overcome with perseverance

[154] Of course this fear of the unbearably depressing subject of the human condition can also cause some people to try to put the blocks to confronting the human condition back in place by contriving all manner of false arguments against the fully explanatory and accountable and thus true explanations of the human condition that are being put forward. Indeed, what is being presented can cause such an angry response that

those supporting the fully accountable, genuinely redeeming and relieving, human-race-saving, biological explanation of the human condition are <u>viciously attacked and persecuted</u>!

Angrily attacking the truth about the human condition

[155] <u>Absolutely amazingly, everything that has just been said about how fearful the human race has been of the human condition and about all the defensive responses that the arrival of its explanation initially causes, was fully anticipated by that absolutely astonishingly truthful and thus effective thinking Greek philosopher, Plato</u>. Plato was such a magnificent philosopher (philosophy being the study of **'the truths underlying all reality'** (*Macquarie Dictionary*, 3rd edn, 1998)) that Alfred North (A.N.) Whitehead, himself one of the most highly regarded philosophers of the twentieth century, described the history of philosophy as being merely **'a series of footnotes to Plato'** (*Process and Reality [Gifford Lectures Delivered in the University of Edinburgh During the Session 1927-28]*, 1979, p.39 of 413). <u>So, way back in the Golden Age of Greece, some 360 years BC, this is what Plato wrote in *The Republic*</u>: **'I want you to go on to picture <u>the enlightenment or ignorance of our human conditions</u>** [and this is the earliest use of the term 'human condition' that I have seen] **somewhat as follows. Imagine an underground chamber, like a cave with an entrance open to the daylight and running a long way underground. <u>In this chamber are men who have been prisoners there</u>'** (tr. H.D.P. Lee, 1955, 514; or you can view all these quotes from *The Republic* that appear in this and the next two paragraphs highlighted where they actually appear in *The Republic* at www.wtmsources.com/227). Plato described how the cave's exit is blocked by a **'fire'** that **'corresponds...to the power of the sun'**, which the cave prisoners have to hide from because its searing,

'painful' 'light' would make 'visible' the unbearably depressing issue of 'the imperfections of human life' (516-517). Fearing such self-confrontation, the cave prisoners have to 'take refuge' 'a long way underground' in the dark 'cave' where there are only some 'shadows thrown by the fire' that represent a 'mere illusion' of the 'real' world outside the cave (515), which are all the human-condition-avoiding, dishonest, so-called 'explanations' that Reductionist, Mechanistic scientists have been giving us for our behaviour (see the earlier explanations of Reductionist, Mechanistic science's human-condition-avoiding strategy in paragraph 36). The allegory makes clear that while 'the sun...makes the things we see visible' (509), such that without it we can only 'see dimly and appear to be almost blind' (508), having to hide in the 'cave' of 'illusion' and endure 'almost blind' alienation has been infinitely preferable to facing the 'painful', depressing issue of 'our [seemingly imperfect] human condition'.

Computer graphic by James Press © 2018 Fedmex Pty Ltd

[156] And, with regard to the problem of the 'Deaf Effect' response the 'cave' 'prisoners' would have to reading or hearing about the human condition, Plato then described what occurs when, as summarised in the *Encarta Encyclopedia*, someone 'escapes from the cave into the light of day' and 'sees for the first time the real world and returns to the cave' to help the cave prisoners 'Escape into the sun-filled setting outside the cave [which] symbolizes the transition to the real world...which is the proper object of knowledge' (written by Prof. Robert M. Baird, 'Plato'; see www.wtmsources.com/101). Plato wrote that 'it would hurt his [the cave's prisoner's] eyes and he would turn back and take refuge in the things which he could see [take refuge in all the dishonest, illusionary

explanations for human behaviour that we have become accustomed to from human-condition-avoiding, Reductionist, Mechanistic science], **which he would think really far clearer than the things being shown him. And if he were forcibly dragged up the steep and rocky ascent** [out of the cave of denial] **and not let go till he had been dragged out into the sunlight** [shown the truthful, **real** description of **our human condition**], **the process would be a painful one, to which he would much object, and when he emerged into the light his eyes would be so overwhelmed by the brightness of it that he wouldn't be able to see a single one of the things he was now told were real.'** Significantly, Plato then added, **'Certainly not at first. Because he would need to grow accustomed to the light before he could see things in the world outside the cave'** (*The Republic*, 515-516). Yes, reading and listening to discussion of the human condition can **'at first'** cause an extreme Deaf Effect where you are not **'able to see a single one of the things…[you are] now told were real'**, but that deafness can be overcome by patiently becoming **'accustomed to the light'**, persevering with reading, watching and listening to the explanation of the human condition.

Image by James Press © 2017 Fedmex Pty Ltd

[157] Plato went on to also warn that when understanding of the human condition eventually arrives it will not only cause a 'Deaf Effect', it will also cause an extremely defensive and angry response in some—writing

that some of the **'cave' 'prisoners' 'would say that his** [the person who attempts to bring understanding to the human condition] **visit to the upper world had ruined his sight** [they would treat him as mad, which is how all denial-free, truthful thinkers we have termed 'prophets' have been treated throughout history—as it says in the Bible, **'was there ever a prophet your fathers did not persecute'** (Acts 7:52)]**, and that the ascent** [out of the cave] **was not worth even attempting** [which is another comment typically made by psychologically exhausted people who want to project their insecurity onto the world and stop the completion of humanity's heroic search for knowledge]. **And if anyone tried to release them and lead them up, they would kill him if they could lay hands on him'** (ibid. 517)!

[158] Thankfully, we live in more civilised times, but we in the WTM have endured years of this vicious, try-to-**'kill him'**-type persecution, which I will describe further on in Part 2.12.

[159] What an absolutely extraordinarily sound and thus truthful and thus effective, denial-free thinker or prophet Plato was!

Part 2.8 Resignation

[160] Since we are not born practising denial of our 2-million-year, soul-corrupted, innocence-destroyed condition, the question arises when did humans individually take up this practice of denial? The answer is that for generation after generation, almost everyone during their early adolescence had no choice but to resign themselves to living in determined denial of that unbearably depressing subject of humanity's soul-corrupted human condition, and of their own soul-corrupted condition as a result of their encounters with the soul-corrupted condition in others while they were growing up. This extremely torturous process of 'Resignation' to living in denial of the human condition that virtually all adolescents have, up until now, had to agonisingly go through is explained in chapter 2:2 of *FREEDOM*, and slightly more fully in Freedom Essay 30—and this historically denied process of Resignation is so important to understand that I strongly urge readers of this book to read that essay if they haven't already done so. And yes, the process of Resignation is another very important truth that we couldn't admit while we couldn't explain our corrupted human condition.

[161] As I will explain more fully in paragraphs 194-195, the reason I have said 'almost' and 'virtually' everyone had no choice but to resign to living in denial of the human condition is because there have always been rare individuals who were fortunate enough to have sufficiently escaped encountering the soul-corrupted state of the world during their upbringing to not be terrified by the issue of the human condition and so not have to resign to living in denial of the human condition—see for example the description of the unresigned prophet Noah in the Biblical story of Noah's Ark and The Flood further on in paragraph 176.

[162] The next four paragraphs (which are mostly taken from that Freedom Essay 30 about Resignation) give an insight into how horrifically agonising the process of Resignation to living in denial of the human condition has been for young adolescents—especially in a world of resigned adults who haven't acknowledged what they are going through.

[163] The English rock band the Beatles' 1970 song *Let It Be*—consistently voted one of the most popular songs of the twentieth century—is actually an anthem to this adjustment that adolescents have historically had to make when confronted with the unbearable **'hour of darkness'** that came from grappling with the issue of all **'the broken hearted people living in the world'**, to **'let it be' 'until tomorrow'** when **'there will be an answer'** (Lennon/McCartney) to all the wrongness in the world, which is the issue of the human condition. So, again, when the great English poet Gerard Manley Hopkins wrote about the unbearably depressing subject of the human condition in his aptly titled 1885 poem *No Worst, There Is None*, his words **'O the mind, mind has mountains; cliffs of fall, frightful, sheer, no-man-fathomed'** did not exaggerate the depth of depression humans faced if they allowed their minds to think about the human condition while it was still to be **'fathomed'**/understood/**'answer**[ed]**'**. Yes, the situation for almost all young adolescents has been that when, in **'my hour of darkness'**, **'Mother Mary comes to me, speaking words of wisdom, let it be, let it be'**—accept the adults' **'wisdom'** and don't allow your mind to go there!

Catherine Yeulet/iStockphoto; yamasan/ AdobeStock; Al Troin/AdobeStock

[164]Of course adolescents haven't wanted to resign to living in denial of the human condition because it meant blocking out all memory of the innocent, soulful, true world and adopting a completely dishonest, superficial and artificial, effectively dead existence. Although rarely shared, adolescents in the midst of Resignation quite often write excruciatingly honest poetry about their impending fate, such as this heartbreaking Resignation poem by 13-year-old Fiona: **'You will never have a home again… Smiles will never bloom from your heart again, but be fake and you will speak fake words to fake people from your fake soul…From now on pressure, stress, pain and the past can never be forgotten / You have no heart or soul and there are no good memories / Your mind and thoughts rule your body that will hold all things inside it; bottled up, now impossible to be released / You are fake, you will be fake, you will be a supreme actor of happiness but never be happy… You will become like the rest of the world—a divine actor, trying to hide and suppress your fate, pretending it doesn't exist…you spend the rest of life trying to find the meaning of life and confused in its maze'**. Clearly, the price of Resignation is *enormous*, but the alternative for virtually all humans of not resigning has been an *even worse* fate because it meant living with constant suicidal depression.

'You are fake, you will be fake, you will be a supreme actor of happiness but never be happy.' Actors in the Broadway musical *On the Town*

[165] It's little wonder then that the human condition has, as I mentioned earlier in paragraph 14, been described so vehemently as **'the personal unspeakable'** and as **'the black box inside of humans they can't go near'** —*and* why it is so very rare to find a completely honest description by adults of adolescents going through the excruciating process of Resignation where they have to block out from their mind the seemingly inexplicable question of their and the human race's soul-corrupted condition. Having already been through this terrible process of Resignation, most adults simply couldn't allow themselves to recall, recognise and thus empathise with what adolescents were experiencing. And so our young have been alone with their pain, unable to share it with those closest, or with the world at large. All of which makes the following account of a teenager in the midst of Resignation, by the American Pulitzer Prize-winning child psychiatrist Robert Coles, incredibly special: **'I tell of the loneliness many young people feel…It's a loneliness that has to do with a self-imposed judgment of sorts…I remember…a young man of fifteen who engaged in light banter, only to shut down, shake his head, refuse to talk at all when his own life and troubles became the subject at hand. He had stopped going to school…he sat in his room for hours listening to rock music, the door closed…I asked him about his head-shaking behavior: I wondered whom he was thereby addressing. He replied: "No one." I hesitated, gulped a bit as I took a chance: "Not your-self?" He looked right at me now in a sustained stare, for the first time. "Why do you say that?"** [he asked]…**I decided not to answer the question in the manner that I was trained** [basically, **'trained'** in avoiding what the human condition really is]…**Instead, with some unease…I heard myself saying this: "I've been there; I remember being there—remember when I felt I couldn't say a word to anyone"…The young man kept staring at me, didn't speak…When he took out his handkerchief and wiped his eyes, I realized they had begun to fill'** (*The Moral Intelligence of Children,* 1996, pp.143-144 of 218). The boy was in tears because Coles had reached him with *some* recognition and appreciation of what he was wrestling with; Coles had shown *some* honesty about what the boy could see and was struggling with, namely the horror of the utter hypocrisy of human behaviour, including his own—the hypocrisy being that we should be ideally behaved but we aren't.

[166]The middle picture above is particularly revealing of this utter hypocrisy of human behaviour. It is clearly a picture created by an adolescent girl in the midst of Resignation. Having not yet blocked out from her mind the extreme contrast between our species' original cooperative, selfless and loving moral instinctive self or soul (which she, like every other human, is born instinctively aware of and instinctively expecting to encounter) and our present horrifically soul-corrupted, angry, egocentric and alienated condition (symbolised by the green-eyed, snarling wolf drawing she has made of herself), she is still wrestling with the horror of the utter hypocrisy of our horrifically corrupted or 'fallen', soul-devastated condition. Saying **'It's not a phase Mom! This is who I really am!'** is in response to her mother having likely dismissively said, "Look, you'll get over it, it's just the 'puberty blues' stage all adolescents go through", which is how the resigned, human-condition-avoiding adult world, including Reductionist, Mechanistic scientists, have been dishonestly explaining the cripplingly distressed and depressed stage adolescents go through during Resignation—yes, the resigned world has had a dishonest, human-condition-avoiding reason for everything! The truth is that what the adolescent girl is going through has nothing to do with the hormonal changes of puberty, it is to do with trying to face down the truth of the horrifically soul-corrupted state of the human condition in the human race, and in herself as a result of her own encounters with the horrifically corrupted state of the human condition in others while growing up.

[167]R.D. Laing was a very experienced and brilliant psychiatrist, and he perfectly understood what happens when an adolescent goes through Resignation, and its effects, which is apparent in his aforementioned quote (see paragraph 149) where he wrote that **'The condition of alienation, of being asleep, of being unconscious, of being out of one's mind, is the condition of the normal** [resigned to living in denial of the human condition] **man. Society highly values its normal man. It educates children to lose themselves** [become resigned]

and to become absurd, and thus to be normal…we are driving our children mad more effectively than we are genuinely educating them…To adapt to this world the child abdicates its ecstasy…the *ordinary* person is a shrivelled, desiccated fragment of what a person can be. As adults, we have [resigned and] **forgotten most of our childhood, not only its contents but its flavour; as men of the** [resigned] **world, we hardly know of the existence of the inner world** [of our soul]**'.**

Drawing by Jeremy Griffith © 2015 Fedmex Pty Ltd

[168] After what has now been explained about the human condition, resigned adults will, with patience to absorb what has been said, now be able to see that they have so blocked out the issue of the human condition that it is the *real* '**elephant in our living rooms**' that everyone has long recognised existed but could never allow their minds to actually identify. Even though the issue of the human condition is *the* one great outstanding and all-important issue that had to be addressed and solved, it is the one great issue all resigned adults, which is virtually all adults, have pretended doesn't even exist. As R.D. Laing was also quoted as saying earlier in paragraph 149, the '**desperately urgently required project for our time – [is] to explore the inner space and time of consciousness** [explore what our conscious mind *really* thinks about]…[yet] **We are so out of touch with this realm** [living in such fearful denial of the issue of our corrupted human condition] **that many people can now argue seriously that it does not exist. It is very small wonder that it is perilous indeed to explore such a lost realm**'! The extreme fear that resigned adults have of '**explor**[ing] **such a lost realm**',

even though it has now finally been explained and rendered safe to look at, is what causes the initial problem people have of the 'Deaf Effect' that was described earlier.

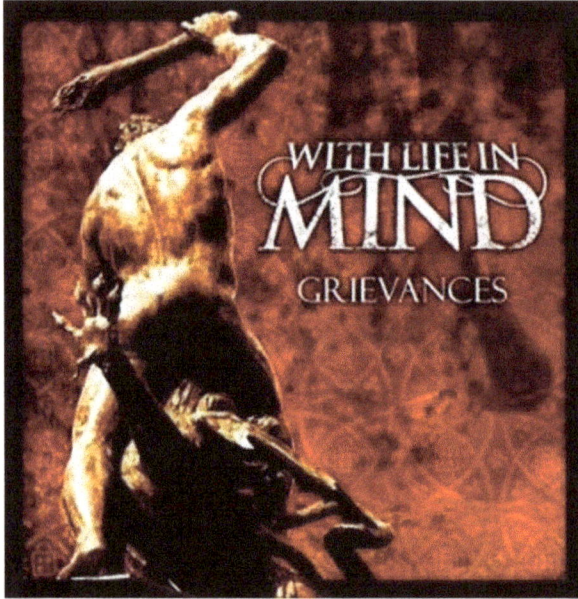

[169] A clear <u>example of a pre-resigned honest mind</u> that can still see the human condition can be found in the lyrics of the American heavy metal band With Life In Mind's 2010 album *Grievances*, which were written by one of the band's members when he was still a young teenager. And look at the cover [above]: the world is mad, there is a person screaming, getting beaten up. 'With Life In Mind' actually should be 'With The Human Condition In Mind'! These are the extraordinarily unresigned, truthful lyrics: **'It scares me to death to think of what I have become…I feel so lost in this world'**, **'<u>Our innocence is lost</u>'**, **'I <u>scream</u> to the sky but my words get lost along the way. I can't express all the hate that's led me here and all the filth that swallows us whole. I don't want to be part of all this insanity. Famine and death. Pestilence and war. A world shrouded in darkness…<u>Fear is driven into our minds everywhere we look</u>'**, **'Trying so hard for a life with such little purpose… Lost in oblivion'**, **'<u>Everything you've been told has been a lie…We've all been asleep since the beginning of time. Why are we so scared to use our minds?</u>'**, **'<u>Keep pretending; soon enough things will crumble to the ground…If they could only see the truth they would coil in disgust</u>'**, **'<u>How do we save ourselves from this misery…So desperate for the answers…We're straining on the last bit of hope</u>**

we have left. No one hears our cries. And no one sees us screaming', 'This is the end.' So that is how honest an unresigned mind can be, and by inference how dishonest a resigned mind is! Saying **'Everything you've been told has been a lie'** emphasises the extent of the dishonest denial in the world of resigned adults, and saying **'So desperate for the answers'** confirms how incredibly precious are the redeeming **'answers'** about our corrupted human condition that have now been found—and are being made available to everyone free of charge on our WTM website.

Munch's *The Scream*, 1893

[170]The earlier mentioned 1893 iconic painting by Edvard Munch, *The Scream*, is a truly exceptional example of the honesty some artists have been capable of—because in this painting Munch has dared to depict the **'scream'** that With Life In Mind said **'no one hears'** or **'sees'**. No one has ever properly explained it, but we now can: *The Scream* is a deservedly famous picture for its portrayal of the human condition—its revealing

honesty is why many pre-resigned teenagers have it pinned up in their rooms. It is an extremely honest and revealing picture because we can see two 'normal' people in the background promenading down the pier—'normal' being totally resigned to pretending everything is fine and the world is as it should be. So they're swanning along the pier saying, we can imagine, 'That's a lovely sunset, should we go down and have an ice cream?', and yet this person in the foreground is screaming, and the whole realm is resonating with the horror of humans' true situation. Yes, these two people, contrasted with the person screaming, is a powerful representation of how dreadful our human situation really is, which people resigned to living in denial of the human condition can't see. The bravery of Munch daring to confront the human condition and reveal the terminal state of alienation in the human race that R.D. Laing described was made clear when Munch said that at one time in his life, **'My condition was verging on madness—it was touch and go'** (*Edvard Munch: Paintings, Sketches, and Studies*, ed. Arne Eggum, 1984, p.236 of 305).

[171] The honesty of Munch's *The Scream* is truly extraordinary; no wonder in announcing its May 2012 auction of the only one of four versions of the *Scream* that was still held privately, Sotheby's auction house described Munch's work as **'the defining image of modernity'** and said they were expecting it to attract one of the highest prices ever for a painting (*The Australian*, 23 Feb. 2012). And indeed it did, selling for almost $US120 million—$US40 million above expectations!

[172] Given how incredibly valuable paintings that only manage to allude to the horror of the human condition are, the WTM's presentations of the actual explanation of the human condition should make them the most valuable commodities on Earth, yet, as I describe later in Part 2.12, instead of financial reward, appreciation and support we are attacked and, at great financial, practical and emotional expense for our small charity, have to fight vicious persecution—yet our generosity and integrity is such that we make all our presentations available free of charge! We humans have been able to cope with the human condition being alluded to, such as in another example, Bob Dylan being given the Nobel Prize for Literature for alluding to the human condition in his songs (see later in paragraph 284), but it takes adjustment time for us humans to cope with the actual truth about the condition. It does need to be said however that

given the serious plight of the human race now, that adjustment time does need to end SOON and at least some serious recognition be given by the establishment to our human-race-saving work at the WTM.

[173] So to be an artist and try to 'cut a window', as it were, through all the pretence and denial to the truth, like Munch did, and Vincent van Gogh also did in his paintings, was incredibly brave. Van Gogh painted light; he taught us to see light. We couldn't see light until Van Gogh painted it. That is true! He was just so honest, and he just kept making himself more and more honest in how he saw the world, until eventually he was able to see the true illuminated beauty of the world! So he painted these amazing pictures and when you look at them you can really see light for the first time, the true brightness of it.

Van Gogh's *The Sower*, 1888   Van Gogh's *Three Sunflowers*, 1888

[174] Van Gogh could paint anything, even these sunflowers in a vase, so they would suddenly come alive. But tragically, in the end, being so honest was unbearable for him and he despaired and went mad and suicided. To be an artist of any sort and dig into and reveal the truth about human life was a tortuous existence—but confronting the truth no longer is because we can now understand the whole immensely, absolutely incredibly heroic journey we humans have been on! So the depth of sensitivity and the extent of creativity we are going to be capable of now will be truly astonishing—but our first task is to disseminate this redeeming and healing understanding of the human condition and stop all the suffering in the world.

[175] All these descriptions about the human condition make it *very* clear just how insecure we humans have been about our corrupted condition and therefore why we have been living such a superficial and artificial existence since Resignation became an almost universal feature of human life around 9,000 BC following the development of agriculture and the sedentary, living-on-top-of-each-other, stressful, alienation-spreading life it created—as the Australian historian Manning Clark said, **'The bush [wilderness] is our source of innocence; the town is where the devil prowls around'** (*The Sydney Morning Herald*, 18 Feb. 1985).

[176] Interestingly, in paragraph 750 of *FREEDOM* and also in Freedom Essay 38, I explain that the Genesis story in the Bible of Noah's Ark and The Flood is actually a metaphorical recognition of how, after the development of agriculture, the human race became so soul-corrupted, so psychologically mind-upset and soul-repressed, that the need to resign to denying the subject of the by then extremely corrupted and thus unbearably depressing issue of the human condition had become an almost universal necessity. So, the story recognises the time when Resignation 'flooded' the world and truth and honesty 'drowned' and only a rare few unresigned truthful thinkers we call 'prophets' survived—like Noah did in his ark because, as it says in Genesis, he **'was a righteous man, blameless among the people of his time, and he walked with God** [he was innocent and sound enough to not have to metaphorically 'drown' and resign to living in denial of Integrative Meaning and any other truth that brings the issue of the human condition into focus, which is most truth]' (Gen. 6:9).

[177] So the Biblical story of Noah's Ark and The Flood doesn't literally refer to a time when a great flood of water destroyed lost ancient civilisations with advanced technologies, as Graham Hancock, Brian Foerster, Randall Carlson, Ben van Kirkwyk and others would have us believe, but to a time when the great flood of neurosis and psychosis and resulting Resignation destroyed humans' ability to think truthfully, effectively and creatively, leaving only a rare few honest thinkers left on Earth—it refers to a time when **'righteous[ness]'** and **'God[liness]'** disappeared! It is a *metaphorical* story, yet people are forever trying to find, and claiming to have found, the actual remains of 'Noah's Ark'!

The Dove Sent Forth From The Ark, Gustave Doré, 1866

[178]One consequence of the emergence of such extreme mind-madness or neurosis and soul-death or psychosis in the human race has been the development of extremely paranoid bewilderment that there is something terribly wrong that we are unable to recognise, which we can now understand is the almost universal denial of the human race's now horrifically corrupted human condition. And an expression of that extremely insecure paranoia has been all manner of conspiracy theories, such as Hancock's and others' belief that the establishment is denying the existence of lost ancient civilisations with advanced technologies. I am currently writing a book about this great paranoia and the conspiracy theories it has given rise to. In that book I explain that there IS a *very* great conspiracy being perpetrated by the 'establishment'—and paranoic conspiracy theorists prey on people's underlying awareness of it. They make a financial

fortune from stories of aliens visiting Earth or lost ancient civilisations with advanced technologies—like the precisely fitted polygonal walls that the engineer Henrique Agostinho posts such interesting YouTube videos about, etc, etc—which the establishment is supposedly supressing. As Christ said, **'Wherever there is a carcass** [the extremely psychologically disorientated]**, there the vultures will gather** [to prey on them]' (Matt. 24:28). But, as I have explained, the very great conspiracy isn't about aliens or lost worlds—it's the human race's outrageously dishonest, almost total denial of our 2-million-year soul-corrupted human condition! On the whole I think YouTubers such as the historian Dr David Miano and history documentary maker Stefan Milo do an excellent job of countering conspiracy theories about lost ancient civilisations with advanced technologies. However, when YouTuber Ben van Kirkwyk made the accusation that the theory he supports—of lost ancient civilisations possessing advanced technologies—is being resisted by the **'mainstream'** because it **'threatens the authority and power of its high priests, our academic establishment'**, and Dr Miano responded by insisting, **'For the last time, there is no such vested interest** [in such a denial]**, it's a myth'** (*Historian reacts to Evidence for Ancient High Technology in Egypt*, 'World of Antiquity' YouTube channel, 22 Nov. 2021), the truth is that both van Kirkwyk and Miano were, in a sense, wrong. There *is* a massive **'vested interest'** by the **'mainstream' 'academic establishment'** in denial—but not of lost ancient civilisations with advanced technologies. Rather, it is denial of our horrifically soul-corrupted human condition. Everyone feels the now absolutely desperate need for truth to replace all the lying that everyone senses is going on in the world, and in that vacuum of bewilderment, lostness and desperate need for the lying to stop, all manner of morbid derangements, conspiracy theories and the like are appearing. The dialogue of one character in the 1991 film *Separate but Equal* accurately recognised the plight of our species when he said, **'Struggling between two worlds; one dead, the other powerless to be born'**—words that echo those of the philosopher Antonio Gramsci: **'The crisis consists precisely in the fact that the old is dying and the new cannot be born; in this interregnum a great variety of morbid symptoms appears'** (*Prison Notebooks*, written during Gramsci's 10-year imprisonment under Mussolini, 1927-1937). Thank goodness a sound and truthful, human-condition-free new world is no longer **'powerless to be born'**, because we have finally found the redeeming understanding of the human condition that enables the human race to be freed and transformed from that dreadful existence.

[179] In earlier times before Resignation and its crippling of our capacity to think and create became universal, people had extraordinary powers of thought, creativity and sensitivity. Look at all that I have been able to explain and create as a result of not having resigned to living in denial of the human condition—my few books of truth have replaced vast libraries of human-condition-avoiding, dishonest, mind-mad and soul-dead books, answers everywhere where there were no answers anywhere.

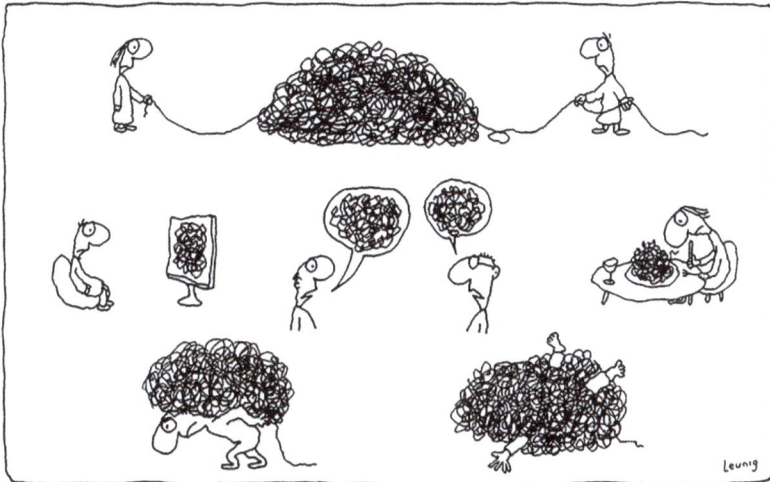

'You read any of Jeremy Griffith's books and there is just truth after truth after truth laid out. He has made sense of and unravelled it all, demystified and worked out every confusion there is. **Michael Leunig** should add another image to his brilliant [9 Aug. 2014] 'impenetrably confusing world' cartoon [above] showing a person lying comfortably on their back in the warm healing sunshine with *FREEDOM* on their lap and all the tangles nicely sorted out in an organised pile beside them' (Ales Flisar, WTM Facebook Group, 7 Jul. 2025).

'In the 1980 film *Blue Lagoon*, the character Richard laments, "I wish a big book with all the answers to every question of the world would drop out of the sky and land in my hands right now. I'd read it till I knew everything!". Well astonishingly Richard, that book that you, and our whole species, has yearned for and dreamed of that answers everything has finally appeared on Earth, its title is *FREEDOM*!!' (Pamela Fairbanks, WTM Facebook Group, 16 Jul. 2025).

[180] Yes, R.D. Laing was right when he wrote (see paragraph 149), '**Our alienation goes to the roots. The realization of this** [truth] **is the essential spring-board for any serious** [truthful, effective, penetrating] **reflection on any aspect of present inter-human life…the** *ordinary* [resigned] **person is a shrivelled, desiccated fragment of what a person can be. As adults, we have forgotten most of our childhood, not only its contents but its flavour; as men of the** [resigned]

world, we hardly know of the existence of the inner world…Our capacity to think…is pitifully limited: our capacity even to see, hear, touch, taste and smell is so shrouded in veils of mystification that an intensive discipline of un-learning is necessary of *anyone* before one can begin to experience the world afresh, with innocence, truth and love…The condition of alienation, of being asleep, of being unconscious, of being out of one's mind, is the condition of the normal [resigned] man…Man cut off from his own mind, cut off equally from his own body – a half-crazed creature in a mad world…between *us* and It [our true selves or soul] there is a veil which is more like fifty feet of solid concrete…To adapt to this world the child abdicates its ecstasy…There is a prophecy in [the Old Testament book of] Amos that there will be a time when there will be a famine in the land, "not a famine for bread, nor a thirst for water, but of *hearing* the words of the Lord [words of truthfulness]." That time has now come to pass. It is the present age', and 'We are dead, but think we are alive. We are asleep, but think we are awake. We are dreaming, but take our dreams to be reality. We are the halt, lame, blind, deaf, the sick. But we are doubly unconscious. We are *so* ill that we no longer feel ill, as in many terminal illnesses. We are mad, but have no insight [into the fact of our madness].'

[181] A further example of an unresigned, immensely truthful and creative mind that is free from the **'fifty feet of solid concrete' 'between *us* and'** our soul, and whose **'capacity even to see, hear, touch, taste and smell'** is not **'pitifully limited'**, was Imhotep, the incredibly creative, clearly un-resigned, mastermind architect of the great pyramids in ancient Egypt. The following are some extracts from the brilliant art historian Lord Kenneth Clark's wonderfully illuminating 1975 BBC documentary *In the Beginning*—note that since this documentary was made in 1975 there may be some archaeological interpretations in these extracts that more con-temporary archaeologists might contest (such as who were the first stone builders in the world), however the absolutely magnificent capabilities of the creators of the pyramids that were built during the Old Kingdom of Egypt (c.2,700-2,200 BC) comes through in Lord Clark's commentary: '[Pharoah] Djoser built…the earliest pyramid. This is Saqqara, to my mind, one of the sacred places of the world, the real birthplace of civilisation. The first thing that strikes one about Saqqara is its lightness and clarity. It's built of an exquisite, creamy, golden limestone. It's the earliest stone building in the world. The cutting is of such perfection, never equalled in Greece or Rome…The love of perfection could not go further…What a refined, sophisticated style. Very far from our idea of what ruins or primitive architecture should be, very far from

Stonehenge, which was built considerably later…One is completely baffled by the extreme sophistication and the incalculable purpose of this exquisite architecture. Combination of measuring mind and responsive hand cannot go further. The date? 2,770 BC…The man who dominates Saqqara is not King Djoser, but a mysterious haunting character named Imhotep. According to Egyptian legend, he was a sort of Prometheus, the first student of medicine, the first student of geometry, the first architect…Imhotep may have been, no must have been, the first man to build in stone. No doubt he was a pioneer, and a universal man of almost legendary stature' (see www.wtmsources.com/307).

Part of the Saqqara mortuary complex designed by Imhotep

[182]The above is a photo of part of the Saqqara mortuary complex designed by Imhotep that reveals the fabulous unadorned, unembellished, egocentricity-free, madness-absent, sound cleanliness and purity of the **'lightness and clarity'** of Imhotep's creativity. Most wonderfully, no longer do we have to be **'completely baffled by the extreme sophistication and the incalculable purpose of this exquisite architecture'** because we can now understand that *it wasn't 'lost ancient civilisations with advanced technologies' but lost innocence* that explains the extraordinary ingenuity and creativity of early civilisations like the ancient Egyptians. And most wonderfully, no longer does the human race have to suffer in this current state of lost innocence, which is the horrifically mind-and-soul-crippled state that R.D. Laing, Christ and the band With Life In Mind described

so truthfully, because, with redeeming understanding of the human condition, the whole human race can be transformed back to soundness, sanity, sensitivity, ingenuity and creativity—that fabulous transformation of the human race will be described next in Part 2.9.

[183]One especially exciting and relieving change that happens now that we can be honest about our corrupted condition is that the whole of anthropology can be reinterpreted truthfully. As explained in paragraph 95, instead of our ancestors being brutal savages aggressively competing with each other to reproduce their genes like other animals, which supposedly gave us brutal competitive, selfish and aggressive instincts that our conscious mind has to always try to restrain, our ancestry is revealed to be the opposite. Some 2 million years ago we began our heroic but horrifically upsetting journey to find conscious understanding of ourselves—and we did so in an all-loving and all-sensitive state. That original state is beautifully captured in Vanessa Woods's description of bonobo behaviour (in paragraph 47): **'Bonobo love is like a laser beam. They stop. They stare at you as though they have been waiting their whole lives for you to walk into their jungle. And then they love you with such helpless abandon that you love them back. You have to love them back.'** It wasn't until the extremely upset-escalating advent of agriculture around 9,000 BC that upset became so great that virtually everyone resigned to living in fearful denial of the issue of the human condition—the result being the **'fifty feet of solid concrete' 'between us and' 'our true selves'** or soul and **'Our capacity to think** [becoming]**...pitifully limited'**, to use R.D. Laing's description. It is now thought that the ruins at Göbekli Tepe—dating to around 9,500 BC and discovered in 1994 on a hilltop in south-east Turkey—was not a residential settlement but a gathering and ceremonial place for people in the region whose way of living was in transition from hunter-gathering to farming. It is thought that around 500 people—seemingly peacefully—came and worked together to build the structures there that feature large 5-metre tall T-shaped pillars (opposite). They didn't know of the wheel, and while they had stone vessels, they had yet to develop pottery, and they carved the pillars out of the local limestone using flint tools, and also made a mortar from the limestone for the floors. These T-shaped pillars that surround each enclosure at the site have belt and arm markings around their waist indicating they represent people, so they very likely represent ancestors. So the inference is that these people who built Göbekli Tepe weren't so upset that they had started warring with each other, which is the extreme upset that close-living agriculture

led to, yet they were no longer innocent enough, and thus free enough of insecurity about their corrupted condition, to not feel the need for the reassuring and comforting presence of their ancestors to counter the utter loneliness of being condemned as evil angry, egocentric and alienated beings. Loneliness of soul had become apparent but not yet ferociously psychologically upset warfare.

Göbekli Tepe in southern Turkey; c.9,500 BC

[184]This is something else I want to explain. As I briefly mentioned in paragraph 41, and explain more fully in chapter 5 of *FREEDOM*, nurturing is what created our moral soul, which means nurturing was the priority throughout our species' early development. As briefly mentioned in paragraph 135, and explained much more fully in paragraphs 769-770 of *FREEDOM*, at 2 million years ago when humans became conscious this female-focused matriarchal situation changed to a male-focused patriarchal situation because, as group protectors, it was men who had to take on the corrupting job of championing our conscious thinking self or ego over the threat to our species of the ignorance of our instinctive self or soul—but this transition wasn't easy. As I explain in chapter 8:11B of *FREEDOM*, women are soul sympathetic, not ego sympathetic, so sooner or later 'ego sympathy', support for our species' upsetting battle to find knowledge, ultimately for understanding of ourselves, which was men's responsibility, had to take over. So the well-established, strong-willed female matriarchy didn't give in to patriarchy for a long time. It took millennia before men were able to contain women's expressed lack of empathy for men's all-important but extremely upsetting battle to defy the ignorance of our

instinctive self—a lack of empathy that was evident in Leunig's Garden of Eden cartoon in paragraph 133, and referred to in paragraphs 135 and 54. Indeed, women were seemingly treated as goddesses in central Europe during the Upper Paleolithic (50,000 to 10,000 BC) and Neolithic (10,000 to c.4,000 BC) periods, as evidenced by the many so-called 'Goddess' or 'Venus' figurines, such as the one pictured below from Çatalhöyük, which is also a Neolithic ruin in southern Turkey some 700 kilometres from Göbekli Tepe. Çatalhöyük is dated from approximately 7,500 to 5,600 BC, which is somewhat more recent than Göbekli Tepe, and unlike Göbekli Tepe is thought to be an agriculture-based residential settlement. As I explain in paragraph 810 of *FREEDOM* about this figurine, we can see from the *extremely* regal stature of the very well-nourished figure seated on her throne of cheetahs just how revered and in control of their societies such strong women must have been.

'Goddess' statue, Çatalhöyük, southern Turkey, c.6,000 BC

Extraordinarily empathetic drawings of animals in
the Chauvet Cave in southern France, c.30,000 BC

[185] Interestingly, anthropologists don't know why the people at Göbekli
Tepe avoided putting faces on the head pieces of their pillars but I think
with our ability to admit the truth of our lost state of innocence we can
understand why. To engrave the faces of beloved ancestors would have
required depicting something of their tortured human condition when what
the people sought was just their comforting presence. I wrote about how
our soul couldn't and didn't want to draw our alienated faces in *FREEDOM*
in paragraphs 831-834. I point out there that the fabulously sensitive and
empathetic 30,000 BC drawings of our early ancestors' very special friends,
the animals at the Chauvet Cave in France (above), contain virtually no
drawings of humans, and none of human faces. As the author Barbara
Ehrenreich noted, **'If the Paleolithic cave painters could create such perfectly
naturalistic animals, why not give us a glimpse of the painters themselves? Almost
as strange as the absence of human images in the caves is the low level of scientific
interest in their absence'** ('The Humanoid Stain', *The Baffler*, No.48, Nov. 2019). I think that
even then the facial expressions of humans were so alienated, so devoid of
the innocence that our faces must have once exhibited, that our instinctive
self or soul couldn't relate to it; it couldn't, and perhaps didn't want to,
draw humans. Even the plentiful goddess figurines typically don't show
facial expressions and only have a blob for a head, with the one shown
opposite being a rare exception, and that exception is from less innocent

and thus less sensitive times. As R.D. Laing said, we were becoming '**a shrivelled, desiccated fragment of what a person can be**'. Yes, we were far more honest back then about our corrupted condition, especially since we hadn't yet resigned to where '**our innocence is lost**' and '**everything you've been told has been a lie**' as the band With Life In Mind sang (paragraph 169), and we had '**darkened…[our] hearts**' and '**all dwell among those who lie**' as Christ said (paragraph 152). I haven't at all been trained to be an artist but, as I have discovered with my writing, if I let my instinctive soul express itself, it is capable of being extraordinarily truthful and empathetic, hence I can sometimes, when my soul feels like expressing itself, create astonishingly empathetic drawings, like my drawing of Christ's mother with Christ as a baby in paragraph 54, or my drawing of Christ in paragraph 152, and in my many other drawings. Truly, as the very great English poet William Wordsworth wrote, '**trailing clouds of glory do we come, From God** [the integrated, loving, all-sensitive state], **who is our home**' (*Intimations of Immortality from Recollections of Early Childhood*, 1807).

[186] The situation at Göbekli Tepe is very similar to what happened in Britain around 3,000 BC where the early Britons were still sufficiently innocent, peaceful and egalitarian that they were able to gather together from all over the British Isles for ceremonies at Stonehenge in southern England. In fact, it has been revealed that the central alter stone was brought all the way from the Orkneys in northern Scotland. Possibly, when anthropology becomes capable of honesty, the circle of huge stones at Stonehenge will also be recognised as primarily representations of soul-and-loneliness-comforting ancestors. I should point out that (as I explain in paragraph 32 of *Death by Dogma* where I talk about ancestor worship) when upset became really extreme following the development of agriculture, we could hardly live with ourselves let alone each other, and when this happened our original love for each other very often became a case of being antagonistic towards each other. (As I explain in paragraphs 906-908 of *FREEDOM*, the Genesis story in the Bible of the struggle between the settled farmer Cain and the nomadic Abel is a recognition of the appearance of this antagonism. Note the incredibly honest and thus insightful anthropological thinking in earlier times when Moses, who wrote Genesis, gave the descriptions in Genesis of the fundamental events in human history of Adam and Eve taking the fruit from the tree of knowledge and being banished from the Garden of Eden

of original innocence, i.e. becoming conscious and as a result upset; and of Cain and Abel to describe the upsetting effects of settled agricultural life on our original innocent soul; and of Noah's Ark and The Flood to describe when humanity was psychologically drowned by Resignation to living in truth-and-soul-destroying denial of the human condition!) And so the more we stopped being fond of each other, the more we stopped wanting to remember our not-so-lovable 'loved ones'. But before love died like this, we did so love each other that we didn't want to let their memory go when they died, and so in those more innocent, less upset times, adoration of our ancestors did play a very important part in our lives. I should also explain, as I do in paragraphs 909-915 of *FREEDOM*, that once agriculture was underway and upset began increasing at a very rapid rate, it wasn't long before it became an 'arms race' of who could become toughest and more brutal and ruthless first, the winners of which in Europe were the Indo-European speaking Yamnaya from the Caspian steppe who wiped out most of the more innocent peoples in Europe. As described in paragraphs 1036-1037 of *FREEDOM*, a descendant of some of the more innocent living in the remote mountains of Wales who survived this rapid development of upset wrote this book. Again, thank goodness all humans can return to peace and togetherness now!

[187] To return to describing Resignation, now that we can explain and defend our horrifically soul-corrupted condition we can at last acknowledge the agonising process adolescents have, until now, had to go through of resigning themselves to living in fearful denial of the unbearably depressing subject of our 'fallen', innocence-destroyed, soul-corrupted human condition. For example, we can at last understand the great Spanish artist Francisco Goya's famous, yet-never-properly-understood-until-now, 1797 etching that he titled *The sleep of reason brings forth monsters* that is reproduced overleaf. 'Why should **'the sleep of reason bring…forth monsters'**; has this poor gentleman suffered a terrible financial loss that he can't bear thinking about; what great tragedy has befallen him'—resigned, human-condition-avoiding minds wonder. No, what Goya graphically depicted is how depressing it has been for us humans to drop our mental guard and think deeply about our horrifically soul-corrupted condition. In the picture we see 'bats from hell'—*extremely* serious worries— tormenting the person's mind!

Goya's *The sleep of reason brings forth monsters*, 1796-1797

[188] **'The sleep of reason brings forth monsters'** because trying to think about the human condition has led to thoughts that are both monstrously large in size, and monstrous in that they are about the possibility that we *are* monsters. Yes, we can understand *very well* why, up until now, we've had to have the attitude described in the Irish rock band U2's 1997 song *Staring At The Sun*: **'It's been a long hot summer, let's get under cover, don't try too hard to think, don't think at all. I'm not the only one staring at the sun, afraid of what you'd find if you take a look inside. Not just deaf and dumb, I'm staring at the sun, not the only one who's happy to go blind.'** The sun is clearly the unbearably confronting and exposing, glaring light of truth about

our corrupted condition that has at last been explained and made safe to confront!

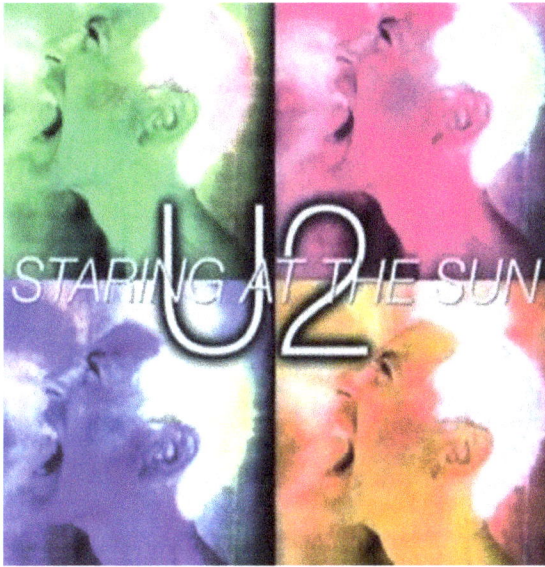

[189] It is true that up until now almost any thinking brought the resigned adult mind into contact with the unbearably depressing issue of our diabolically soul-corrupted, competitive, selfish and aggressive, seemingly-utterly-evil condition. Virtually any thinking was **'too hard'** so best **'don't think at all'**—'Oh, that's a lovely sunset, ooohh, I wonder why I'm not lovable!'; 'Pass me the salt, yes, have I told you about *me*, okay, I know, I'm so focused on me, I'm a fucking insecure, egomaniac wreck of a person, get off my case!'

[190] The absolute reality for virtually everyone has been that trying to confront the issue of our corrupted human condition while we haven't been able to explain it has been such a **'shattering experience'**, as Jung described it (see paragraph 126), that virtually any deeper meaningful thinking would bring your mind into contact with that intolerably, suicidally depressing issue! The Australian comedian Rod Quantock certainly wasn't joking when he said, **'Thinking can get you into terrible downwards spirals of doubt'** ('Sayings of the Week', *The Sydney Morning Herald*, 5 Jul. 1986). And the great truthful-thinking French Nobel Laureate for Literature, Albert Camus, certainly wasn't overstating the difficulty of thinking when he

said, **'Beginning to think is beginning to be undermined'** (*The Myth of Sisyphus*, 1942); nor was another Nobel Prize winner in Literature, the Welshman Bertrand Russell, when he said, **'Many people would sooner die than think'** (Antony Flew, *Thinking About Thinking*, 1975, p.5 of 127), and nor was yet another Nobel Laureate for Literature, the American-English poet T.S. Eliot, when he wrote that **'human kind cannot bear very much reality'** (*Burnt Norton*, 1936). And what did R.D. Laing say (see paragraph 149), he said **'Our capacity to think... is pitifully limited'**. And the great Russian novelist Fyodor Dostoevsky was essentially referring to our inability to engage with the issue of our corrupted condition when he wrote that **'Nothing in this world is harder than speaking the truth'** (*Crime and Punishment*, 1866, ch.4). Also, that very great Greek philosopher Plato recognised the problem of our species' core insecurity about facing exposure to the issue of our corrupted condition when he is widely attributed as having said, **'We can easily forgive a child who is afraid of the dark; the real tragedy of life is when men are afraid of the light.'**

[191] **'The sleep of reason'**, letting down our mental guard and allowing ourselves to think at a deep level, most *certainly* has brought **'forth monsters'**, and so the human race *certainly* has had to live in denial of our 2-million-year soul-corrupted condition while we couldn't explain it! While we humans will readily focus on a safely sectioned-off area of inquiry or activity, such as solving a maths equation, or mastering a computer problem, or remembering an endless stream of meaningless facts for an exam like Queen Isabella the 5th married King Fred the 12th in 1522 and together they fought The War Of The Itchy Bums in 1591, or glamorising our wardrobe, or purchasing a flash car, or even sending a man to the Moon, we won't go beyond those safe limits and risk encountering anything to do with the unbearably depressing subject of our and the human race's corrupted human condition. As a result, there is an *immense* disparity between our superficial outer world and the miles-deep inner world that we won't go near. R.D. Laing's description of there being **'fifty feet of solid concrete'** **'between us and...our true selves** [or soul]' is so true! And yes, this means the real frontier is not outer space but inner space—which is what, in my book *Death by Dogma*, I told Elon Musk, that promoter of outer space exploration and hugely successful innovator and champion of free speech. Hopefully someone will show him the comment in that book.

[192] Our species' extraordinary, indeed mad, situation was well sum-marised by the American General Omar N. Bradley when he said, '**The world has achieved brilliance...without conscience. Ours is a world of nuclear giants and ethical infants**'! (Armistice Day Address, 10 Nov. 1948, *Collected Writings of General Omar N. Bradley*, Vol.1).

[193] Given how fearful of the issue of our corrupted human condition we have been, it is not surprising that we have hardly been able to think truthfully and effectively at all, and therefore why we have hardly made any real headway in understanding ourselves and ending our situation of being '**ethical infants**'. The great Russian philosopher Nikolai Berdyaev recognised not only how difficult confronting the human condition has been, but also how finding knowledge depends on confronting not avoiding the human condition, when he wrote that '**Knowledge requires great daring. It means victory over ancient, primeval terror**...it must also be said of knowledge that it is bitter, and there is no escaping that bitterness... Particularly bitter is moral knowledge, the knowledge of good and evil. But the bitterness is due to the fallen state of the world...There is a deadly pain in the very distinction of good and evil, of the valuable and the worthless**' (*The Destiny of Man*, 1931; tr. N. Duddington, 1960, pp.14-15 of 310). Yes, trying to confront the issue of our corrupted or '**fallen**' condition without the explanation for it only left us with the '**bitter**' '**ancient, primeval terror**' that we humans have had to endure of thinking we must be '**evil**', '**worthless**' monsters — and yet confronting it *is* what was '**require**[d]' to find '**knowledge**', ultimately self-knowledge, the redeeming understanding of our '**good and evil**'-stricken human condition.

[194] Berdyaev has made the point here that while virtually everyone has lived in fearful terror of the human condition, ultimately it *had* to be confronted to be solved. And to do that was going to require very rare individuals who were fortunate enough to have sufficiently escaped encountering the soul-corrupted state of the world during their upbringing to not be terrified by the issue of the human condition and, as a result, to not have to resign to living in denial of it — like Noah was able to avoid doing. It is this unresigned ability to confront and think truthfully and thus effectively about the human condition that the South African philosopher Sir Laurens van der Post and I possessed, and as a result of the confirmation of my truthful thinking that I derived from the honesty

of Sir Laurens's writings, and from the honest thinking of others like him, I was able to find **'the knowledge of good and evil'** explanation of the human condition.

[195] You can read more about Sir Laurens's life in Freedom Essay 51, and in my book *How Laurens van der Post Saved The World*—and also about my life in my biography, and also where I talk about my life in *How Laurens van der Post Saved The World*, and also in Part 7 of another of my books, *Don't Stand In The Way, For The Times Are A-Changin'*, where I describe 'The Golden Thread' of sound thinkers who helped preserve my soul so that I could go on and think sufficiently truthfully to solve the human condition. Those Golden Thread members and their effects are Plato's soul-preserving, **'philosopher kings'**-producing attitude for education, which inspired Dr Kurt Hahn to create the soul-preserving Gordonstoun school in Scotland, which in turn inspired Sir James Darling to create the soul-preserving Geelong Grammar School in Australia that I was so exceptionally fortunate to attend (although I should say that the school lost its Darling-inspired focus for education as soon as Sir James retired). Then there was Sir Laurens van der Post who was one of Sir James's two favourite authors. And then there was my good friend Steve van Hemert, whose wild, larger-than-life, expose-the-world-of-denial-for-the-crap-it-really-is carryings-on were so relieving for me to be part of as a young man. And, above all of course, in terms of the soul-preserving influence in my life, there was the exceptionally unconditionally nurturing love I received from my parents Norman and Jill Griffith. This was 'The Golden Thread' of influences that, as R.D. Laing said (see paragraph 149), allowed me to **'leave the formation [that]…is itself off course** [stay away from the resigned world of denial of the human condition]' and **'get 'on course"** and solve the human condition. That Part 7 about 'The Golden Thread' in *Don't Stand In The Way, For The Times Are A-Changin'* is very revealing of how I was able to confront and solve the human condition. It does make understandable why Professor Anthony Barnett, zoologist, author and broadcaster of a popular science program in Australia, said to me back in 1983 that **'you are being very arrogant to think you can answer questions on this scale** [confront the human condition as he recognised I was doing]. **In the whole of written history there are only two or three people who have been able to think on this scale about the human condition'** (from a recorded interview with Jeremy Griffith, 15 Jan. 1983).

Our present dystopia caused by our fear of the scorching 'fire' of the issue of the human condition, and the quenching of that 'fire' and resulting liberation and transformation of humanity from having to live with the agony and horror of the human condition

[196]This is a drawing I have done (which also appears on this book's cover) to depict our situation now. On the left is our present dying world which has become a dystopian wasteland where human progress has foundered and science has lost all credibility due to dishonest Reductionist, Mechanistic science's fearful avoidance of the 'fire' of the human condition. On the right of the fire are the only people able to confront and think truthfully about the human condition—the very rare human-condition-confronting truthful thinkers or 'prophets', symbolised by those with halos around their heads. The circular glow of light or 'halo' often drawn around the heads of prophets was used to indicate their soulful purity, innocence, soundness and holiness; indeed, the word 'holy', so often used to describe prophets, has the same origins as the Saxon word 'whole', which means 'well, entire, intact', and is thus a recognition of the prophets' wholeness or soundness or lack of separation or alienation from our species' sound, innocent, all-loving and all-sensitive original instinctive self or soul. Such soundness was needed to synthesise the redeeming explanation of humans' corrupted condition, which is symbolised by a prophet putting the fire out with a bucket of water/truth. Then, as depicted on the right, with the 'fire' finally put out, in other words with our fear of the truth of our corrupted condition at last extinguished, the

whole human race can finally progress to a fabulous 'sun'/truth-drenched, transformed world free of the agony and horror of the human condition!

[197] I should mention that it is not surprising that throughout history fire has been used as a metaphor for the great fear humans have had of the unbearably depressing—'incinerating'—issue of the human condition. For example, in religious texts, fire appears as a metaphor for the integrative, Godly ideals of life whose condemning, scorching glare humans have had to hide from. For example, in the Zoroastrian religion, **'Fire is the representative of God…His physical manifestation…Fire is bright, always points upward, is always pure'** (Edward Rice, *Eastern Definitions*, 1978, p.138 of 433). In the Bible, Moses's story in Genesis features **'a flaming sword flashing back and forth to guard the way to the tree of life'** (3:24), and the Bible also records the Israelites as saying, **'Let us not hear the voice of the Lord our God nor see this great fire any more, or we will die'** (Deut. 18:16). And, as mentioned in paragraph 155, Plato wrote about the **'fire'** that **'corresponds…to the power of the sun'**, which the **'cave'** prisoners have to hide from because its searing, **'painful' 'light'** would make **'visible'** the unbearably depressing issue of **'the imperfections of human life'**, which is the issue of **'our human condition'**. And in paragraph 131, Blake wrote about **'the fire'** that **'could twist the sinews of thy heart'** with unbearable self-confrontation.

Part 2.9 The fabulous, indescribably wonderful transformation of the human race from having to live with the agony and horror of the human condition to living free of it

[198] This sculpture, *Weight of Thought*, by the Belgian artist Thomas Lerooy captures the unbearable agony of trying to make sense of our seemingly mad human condition. As an article on the Artsology website says, **'The sculpture, with its exaggerated head and diminutive torso, stands as a powerful symbol of the burdens that knowledge and thought can impose on the human spirit'**, and that **'Its themes of mental burden and the search for meaning'**, which **'Lerooy sought to encapsulate in bronze – a material as timeless as the human condition it represents'**, **'continues to resonate with audiences'** ('The Gravity of Intellect: Thomas Lerooy's "Weight of Thought"', 20 Apr. 2024; see www.wtmsources.com/305).

Thomas Lerooy's *Weight of Thought*, 2009

[199] Yes, finding understanding of the human condition and by so doing ending **'the search for meaning'** is what allows everyone to finally lift the **'mental burden'** of not being able to understand **'the human condition'**, and by so doing make sense of human existence. It gives us the **'Understandascope'** Michael Leunig was brilliantly intimating we desperately needed to rehabilitate and transform the human race in the following cartoon. This is the fourth masterpiece from Leunig that has been included in this book; three were included earlier in paragraphs 133, 142 and 179. With Leunig's death in 2024 the world lost a brilliantly brave, honest and insightful, prophetic talent.

Cartoon by Michael Leunig, Melbourne's *The Age* newspaper, 17 Mar. 1984

[200] So how does the '**Understandascope**' of the human condition actually liberate and transform the human race from having to live with the agony and horror of the human condition?

[201] As has now been carefully and fully explained, lacking the *real* defence of the Instinct versus Intellect explanation for our corrupted condition all we humans have had to cope with the unbearably depressing truth of our corrupted condition are the *artificial* defences of attacking any criticism of ourselves, of achieving any positive reinforcement of ourselves that we could find, and of denying and blocking out from our mind the condemning truth of our soul-corrupted condition—our angry, egocentric and alienated ways of coping are what have sustained us. But now that we have the *real* defence for our corrupted condition all those artificial defences of anger, egocentricity and alienation are obviously no longer needed; they are obsoleted and can end—the result of which is the great transformation of the human race from living with the agony and horror of the human condition to living free of it!

[202] While we now have the psychologically relieving understanding to rehabilitate the human race, the return to a completely psychosis and neurosis and alienation-free state will naturally take a number of generations. This is because it usually takes years of psychological healing for

all the childhood hurts to our soul and resulting insecurities, confusions and misunderstandings and their effects to be replaced and repaired with understanding, so it follows that in the case of the *largest* neurosis and psychosis of all of the human condition, that healing process will be a generational process. (See my book *Therapy For The Human Condition*.)

[203] HOWEVER, while the complete psychological rehabilitation of our species will take a number of generations, <u>WHAT IS OF IMMENSE IMPORTANCE AND IS SO SPECTACULARLY WONDERFUL is that every human can immediately *know* that he or she is fundamentally good and not bad, and that this knowledge puts each person in a very powerful position *because it means we can legitimately decide not to live in accordance with all the neurological and psychological upset within us*.</u>

[204] The logic behind making this decision is irrefutable. As I have said, now that the great goal of the whole human journey of conscious thought and enquiry is achieved and we have found understanding of our conflicted and distressed human condition, all the old retaliatory, defensive and insecure behaviours of anger, egocentricity and alienation that we had to employ to cope while we couldn't defend ourselves with understanding are no longer needed. They are obsoleted, brought to an end. <u>In fact, with this knowledge of the human condition now found, it would be an act of *total irresponsibility, indeed madness*, to continue down that old, insecure, defensive, retaliatory and destructive road. The truth is, there is nothing in the way now of every human taking up a magnificent, unburdened, *Transformed* Way of Living where we abandon those old, now obsoleted, ways of living.</u>

[205] And, gloriously, what happens when we give up our old way of living and take up the new way of living that understanding of the human condition has made possible, is we transition from a competitive, selfish and aggressively behaved person to a cooperative, selfless and lovingly behaved person. <u>Even though we are not yet free of the psychologically upset state of our own personal human condition, we *can immediately* have a change of attitude and decide not to live out that upset state that remains within us. The overall effect in our lives is that, despite the upset state of the human condition that still exists within us, we are EFFECTIVELY free from its hold and its influence, which is an absolutely fabulous *transformation* to have made!</u>

[206] Every human *can*, as it were, put the issue of all their mental upsets and soul-corruptions in a 'suitcase', attach a label to it saying 'Everything

in here is now explained and defended', and simply leave that suitcase behind at the entrance to what we in the World Transformation Movement call the Sunshine Highway, and set out unencumbered by all those upset behaviours into a new world that is effectively free of the human condition.

²⁰⁷ What has just been described may at first sound extraordinary and unrealistic but it actually isn't.

²⁰⁸ As I explain in detail in chapter 8 and chapter 9 of *FREEDOM*, and also in my book *Death by Dogma*, humans *are* able to do this, decide to change from living for self to living a more selfless existence—*because we have been doing it throughout our history*. In chapter 8 of *FREEDOM* I describe the heroic journey our species has been on as it attempted to find ways of living, and causes to support, that helped to contain and, in some cases, completely transcend our upset, resigned, self-centred, egocentric, selfish way of living. There was self-discipline where we 'civilised' our upset behaviour by superimposing more ideal principles for living upon our selfish and aggressive psychologically upset way of thinking and responding to life. Then there was the imposed discipline of our upset, where we lived by adhering to specific laws of behaviour, such as Moses's Ten Commandments. Then there was religion, where we deferred to and lived through our support of an exceptional denial-free thinking, truthful, unresigned, sound prophet. And then there was a sequence of increasingly guilt-relieving, pseudo idealistic,

no-psychologically-relieving-understanding-of-the-human-condition-based, false-starts-to-an-ideal-world, attitudes that were adopted, from socialism, to the New Age Movement, to feminism, to the environment or green or climatism movement, to the politically correct movement, to finally the completely guilt-stripped and totally dishonest, 'there-is-no-such-thing-as-truth' postmodern or deconstructionist movement, and its latest manifestation, the outrageously dishonest and deluded Critical Theory-based, 'woke', 'Great Reset', Globalisation, 'No Kings' movement.

[209] Of course, as I emphasised in chapter 9 of *FREEDOM*, the degree to which we abandoned, or left behind, or let go, or repressed, or transcended our selfish and aggressive upset, real self and superimposed upon it or adopted a more selfless and cooperative attitude, varied enormously, but within that spectrum there existed the possibility of *completely* abandoning our upset way of behaving. In Christianity for example, this complete so-called 'conversion' from living for yourself to living for Christ is described as being **'born again'** (John 3:3); as becoming **'a new creation'** where **'the old has gone, the new has come!'** (2 Cor. 5:17). So, just as what happened with Christianity, while resigned humans can't easily change from being psychologically upset—as emphasised, that is a process that will take a few generations—we *can immediately and completely* change our mind's *attitude* from living a selfish, self-preoccupied life and be, as it were, 'born again' to a consider-the-welfare-of-others-above-your-own-welfare, unconditionally selfless, soulful, pre-resigned-like way of thinking and living. While every resigned human naturally becomes extremely habituated to living for the relief of power, fame, fortune and glory, it *is* possible to completely relinquish that way of thinking and living and, in its place, adopt a completely different, unconditionally selfless way of thinking and behaving.

[210] It immediately needs to be emphasised that the Transformed Way of Living is not a religion. While it is true that it is similar to a religion in that it involves completely letting go of or transcending our real, upset self and deferring to another way of living, that is where the similarity ends. The fundamental difference between the Transformed State and a religious conversion is that the Transformed State *is all about knowledge, not faith or dogma*—and knowledge that, after a few generations, has the ability to eliminate all the upset in humans. As is emphasised in *Death by Dogma*, dogma can't heal upset, only understanding can do

that. While religions *were* an incredibly effective means of containing the upset in humans while the search for understanding of that upset condition was being carried out, the Transformed Way of Living, and the World Transformation Movement that promotes it, is, in complete contrast, concerned with what happens *after* that liberating understanding is found, which is the advancement of the human race from a human-condition-afflicted state to a state completely free of that terrible affliction. The Transformed Way of Living, and the psychological amelioration of the human condition that it allows (given the time, the few generations, needed for it to take place), fundamentally changes the human race, taking it from a state of psychologically troubled upset to a state of psychologically secure soundness.

[211]Religions were based on deferring to, and living through support of the *embodiment of the ideals* in the form of the soundness and truth of the unresigned, denial-free-thinking-and-behaving prophet around whom the religion was founded, whereas the Transformed State is based on deferring to and living through support of first-principle-based *understandings of the ideals and of our species' unavoidable historical lack of compliance with those ideals*.

[212]In fact, the Transformed Way of Living, where humans live in support of the understanding of the human condition that can, after only a few generations, eliminate all the psychological upset in the human race, represents the *realisation* of religion's hope and faith that the liberating understanding of the human condition would one day appear—it represents the end of faith and dogma and the beginning of knowing. So this is the end of the need for religion and the beginning of the time of understanding that, in fact, all the great prophets looked forward to.

[213]Yes, religions aren't being threatened by the arrival of dignifying understanding of the human condition—they are being fulfilled. The whole purpose of religion was to be the custodian of the ideal state while the search for the liberating understanding of humans' 'fallen' condition was underway. Buddha, for instance, looked forward to the arrival of the amelioration of the human condition when he said that **'In the future they will every one be Buddhas** [meaning in the future everyone will be free of psychosis] / **And will reach Perfect Enlightenment / In domains in all directions / Each will have the same title** [there will be no more distorting alienation] / **Simultaneously on wisdom-thrones / They will prove the Supreme Wisdom'** (Buddha [Siddartha Gautama] 560–480 BC, *The Lotus Sutra*, ch.9; tr. W.E. Soothill, 1987, p.148

of 275). In the Bible, Moses similarly anticipated a time when we **'will be like God, knowing'** (Gen. 3:5). In his *Lord's Prayer*, Christ instructed us to pray for the time when **'Your** [Godly, integrated, peaceful] **kingdom come, your** [integrative] **will be done on earth as it is in heaven'** (Matt. 6:10 & Luke 11:2). He also looked forward to the time when **'another Counsellor to be with you forever—the Spirit of truth** [the denial-free, truthful, first-principle-based, scientific understanding]**...will teach you all things and will remind you of everything** [all the denial-free truths] **I have said to you'** (John 14:16, 17, 26). He similarly said he looked forward to when, instead of being restricted to **'speaking figuratively'**, we **'will no longer use this kind of language but will** [be able to] **tell you plainly about my Father** [be able to explain the world of Integrative Meaning in denial-free, human-condition-reconciled, compassionate, understandable, rational, first principle, scientific terms]**'** (John 16:25). And again, the same anticipation of our species' liberation from the human condition is expressed in Revelations in the Bible where it states that **'Another book** [will be]**...opened which is the book of life** [the human-condition-explaining and humanity-liberating book]**...**[and] **a new heaven and a new earth** [will appear] **for the first heaven and the first earth** [will have]**...passed away...**[and the dignifying full truth about our condition] **will wipe every tear from...**[our] **eyes. There will be no more death or mourning or crying or pain** [insecurity, suffering or sickness]**, for the old order of things has passed away'** (20:12, 21:1, 4). Yes, as Isaiah hoped, there would come a time when humans **'will beat their swords into ploughshares...Nation**[s] **will...**[not] **train for war any more'** (Isa. 2:4). And what did that truth-saying prophet John Lennon **'imagine'**? A time when the human condition is resolved and **'the world will be as one'**, when there will be **'no heaven** [above us and] **no hell below us'**; when, in essence, there will be a world without the condemning differentiation of good and evil, a world liberated from the insecurity of the human condition and thus the need for religion, where, as Lennon sang, there will be **'Nothing to kill or die for, and no religion too...all the people living life in peace...No need for greed or hunger, a brotherhood of man...all the people sharing all the world'** (*Imagine*, 1971). And, as has already been mentioned, it is the fulfillment of Martin Luther King Jr's **'dream'** of the time when the human race is able to **'allow freedom to ring...from every village and every hamlet, from every state and every city'** because **'all of God's children, black men and white men, Jews and Gentiles, Protestants and Catholics'** can **'join hands and sing'**, **'Free at last! Free at last! Thank God Almighty, we are free at last!'** ('I Have A Dream' speech, 28 Aug. 1963).

[214] Another immense difference between the new Transformed Way of Living and a religion is that in the Transformed Way of Living <u>there is no deity involved, or deference to any one personality; in fact, there is no worship of any kind. And best of all, unlike religion, there is no involvement or emphasis on guilt, because guilt—and the whole notion of 'good and evil'—has been eliminated forever with the reconciling understanding of the human condition.</u>

[215] So while religion and the human-condition-resolved Transformed Way of Living both involve letting go of living through our corrupted self and deferring to something else, that is where the similarity ends.

Graphic by J. Griffith, M. Rowell and G. Salter © 2009 Fedmex Pty Ltd

[216] So yes, as depicted above, with understanding of the human condition at last found we are no longer trapped in Plato's horrible cold, dark cave of denial and delusion, and are free to come out into the warm healing sunshine of the compassionate full truth of the redeeming and dignifying understanding of ourselves. Our insecurity about being corrupted ends. We can finally understand that we are all good and not bad. And what this means is that all our old artificial and superficial dishonest and deluded ways of coping are no longer needed. We are FREE to leave all those fake and in-truth horrid practices behind as finished with. We can finally transition from living in a dreadful state of denial and delusion to living in a wonderfully TRANSFORMED STATE of freedom from all the dishonesty and delusion that made human life so fraudulent, mad,

destructive and chaotic. <u>In fact, all the in-truth extreme shallowness, emptiness and loneliness of that cold and dark dishonest and deluded, immensely destructive existence is able to be replaced by so much relief, happiness and healthiness now that we will find it almost too exciting to bear! We go from living an existence tortured by the horror of the human condition, to an absolutely fabulous life free of that horror.</u>

[217]Later in paragraph 292 I use the example of the immense relief the Christian apostle St Paul experienced when he gave up living for his corrupted condition and instead chose to live through his support of Christ. Well, I will pre-empt that and include here this picture I have drawn of St Paul falling off his donkey and going blind for three days with ecstatic relief. That is how much more relieving it is now to be supporting the actual human-race-transforming understanding of the human condition! As I say in that later paragraph, the relief and happiness that is on offer now is so incredibly great it is like falling off the back of a thousand donkeys and of our eyes being so flooded with tears of happiness we won't be able to see out of our eyes for months!

St Paul falling off his donkey and going blind with ecstatic relief
after he let go of his struggle with the human condition

[218]The following two paintings by the already mentioned prophetic artist and poet William Blake, titled *Cringing in Terror* and *Albion Arose*, <u>powerfully illustrate the absolutely fabulous transformation from having to live with the horror and agony of the human condition, to being able to live free of it in support of the redeeming, reconciling and healing understanding that we now have of that condition.</u>

William Blake's *Cringing in Terror* (c.1794–96) left, and *Albion Arose* (c.1794–96) right. This 1991 coloured impression of *Albion Arose* is by WTM friend Carol Marando.

[219] Basically, we humans can now come back to life from having had to endure living in an effectively dead state. As the descriptions from R.D. Laing, Christ, and the band With Life In Mind made clear, 2 million years of accumulating upset *has* left us in a numb, seared, effectively dead state—but we are finally free now from having to live trapped in Plato's deathly dark, deep, cold cave of denial and delusion. The whole human race can wake up now from what has in truth been a terrible, terrible nightmare. WE ARE FREE TO ALL COME BACK TO LIFE NOW; the great TRANSFORMATION of the whole human race is on! And this is not another deluded, pseudo idealistic, false start to an ideal world, this is the real deal, the human-condition-understood-and-resolved ACTUAL transformation of the human race!

[220] While chapter 9 of *FREEDOM* gives an overall description of how this completely wonderful transformation takes place, my booklet/video *The Great Transformation*, my booklet/video *Sermon On The Beach*, and my Video/Freedom Essay 33, provide a specific description of how to let go and leave the old insecure, have-to-prove-you-are-good-and-not-bad ways of living and adopt the new, human-condition-free Transformed Way Of Living where you can live a fabulously meaningful and joyous life in support of the redeeming understanding of our corrupted condition. Also, the many affirmations, endorsements, ecstatic responses and excited descriptions on our website all bear out how fabulous the new world is that has opened up for the whole human race!

[221] It does need to be emphasised that with humanity's immensely heroic but at the same time immensely upsetting search for the redeeming explanation of our corrupted condition finally over, not only CAN everyone leave the old, now obsoleted, ways of sustaining their sense of self-worth and instead adopt the new Transformed Way Of Living, everyone actually HAS TO leave that old way of living and instead live in support of understanding of the human condition. What makes everyone HAVE TO make this transition is not the dictates from some outside authoritarian regime, or the imposition of a dogma-based, pseudo-idealistic, 'woke' culture, but the imposition of the unassailable logic that the old selfish artificial ways of living are destroying the world and can no longer be endured, and since those ways of living are now no longer needed there is NO JUSTIFICATION AT ALL for continuing to live that way. ADOPTING THE TRANSFORMED WAY OF LIVING IS WHAT EVERYONE CAN AND THEREFORE HAS TO DO NOW TO SAVE HUMANKIND.

[222] I should clarify, as I do in paragraph 1231 of *FREEDOM* and in Freedom Essay 33, that the focus in leaving the old world of artificial reinforcement isn't on giving up your possessions or walking the streets in sackcloth in self-denial and servitude. We're talking about a change of mindset that can have an effect on your priorities, which can affect your choice of possessions and so forth, but the focus isn't on self-deprivation.

[223] I also want to point out, as I do in my book *Death by Dogma* in paragraphs 94-96, we humans have always known that one day we had to mature from living in an insecure, defensive competitive, selfish and aggressive state to living in a secure cooperative, selfless and loving state, but being unable to confront the human condition and by so doing find the understanding that actually makes that transformation possible, out of sheer desperation it was decided to just 'fake it to make it', which is what the Critical-Theory-based, 'woke', 'Great Reset', Globalisation, 'No Kings' movement have been trying to do. While such fakeness was never going to work—as *Death by Dogma* makes clear, only the psychologically redeeming, reconciling and healing understanding of our corrupted condition could relieve us of that condition—it does reveal how intuitively anticipated and needed this great transformation is. Later in paragraphs 289-290 I will include descriptions by the famous podcasters Joe Rogan, Russell Brand, Neil Oliver and Kevin A. Maclean of a great yearning now for a whole new way of living based on **'the sanctity of love and truth'**, as Rogan put it—which the human race now has in the human-condition-solved Transformed Way of Living!

[224] So yes, taking up the Transformed State is all-relieving, all-wonderful and all-important! Join the Sunshine Army on the Sunshine Highway to the World In Sunshine and let's fix the world because at last we really can!

Part 2.10 How the human race is able to cope with 'exposure day' or 'judgment day'

[225] Having found the fully accountable and thus true Instinct vs Intellect explanation of the human condition (as Albert Einstein said, **'truth is what stands the test of experience'** (*Out of My Later Years*, 1950)), the challenge now for the human race is to be able to firstly overcome the Deaf Effect and 'hear' the explanation, and then to adopt the just described fabulously relieving and exciting for the individual, and fabulously human-race-saving, Transformed Way of Living that the explanation makes possible.

[226] What now needs to be looked at is how to cope with the *initial* shock of the sudden exposure of our corrupted condition; cope with what has historically been referred to as 'judgment day'.

[227] The great playwright George Bernard Shaw's observation about breakthroughs in science, that **'All great truths begin as blasphemies'** (*Annajanska*, 1919), most especially applies to the arrival of the explanation of the human condition. Since the all-important issue for biology to solve for there to be a future for the human race of our 'soul-corrupted', 'fallen', 'good and evil'-stricken human condition has historically been the most difficult of all subjects for humans to face, it has to be expected that the arrival of its explanation will initially be the most **'blasphem[ous]'**, and thus the most resisted, of **'all great truths'** in science. While, as I explain in chapters 2:10-2:11 of *FREEDOM*, the renowned Harvard biologist E.O. Wilson's Sociobiology and Multilevel Selection 'savage instincts'-based theories for human behaviour are false and extremely dangerously misleading, I fully agree with Wilson's recognition that **'The human condition is the most important frontier of the natural sciences'** (*Consilience*, 1998, p.298 of 374), and **'There is no grail more elusive or precious in the life of the mind than the key to understanding the human condition'** (*The Social Conquest of Earth*, 2012, p.1). Yes, while **'understanding the human condition'** is the most **'important'**, **'precious in the life of the mind' 'grail'** of science, it is also the most **'elusive' 'grail'**

because it has been the most psychologically confronting of all subjects for humans to engage with.

[228] Since it is well known by psychologists that **'denials fight back with a vengeance when faced with annihilation'**, it is not surprising that this most denied and repressed of all truths of the issue of our soul-corrupted human condition was going to initially meet extreme resistance when the real and true analysis of it appeared. And since there has been no greater **'blasphemy'** for human-condition-avoiding Reductionist, Mechanistic scientists than confronting the human condition, it is not surprising that appreciation and support for this most needed of all scientific break-throughs was going to take time to build in the scientific community. However, while scientists, and virtually the whole human race, have been avoiding the human condition, now that the human condition has been explained and made safe to confront, that avoidance is no longer justified and the wider scientific establishment, along with the general public, *can*, *should* and *needs* to support this breakthrough—and, as I said above in paragraph 221, they need to support it SOON or the human race will perish from terminal levels of psychosis. (The reason I have just said that the 'wider scientific establishment' needs to support this breakthrough is because while there are already ever-increasing numbers of eminent, truth-acknowledging scientists who have appreciated and publicly acknowledged the substance and importance of the breakthrough, such as those mentioned in the commendations section on our website at www.humancondition.com/reviews-commendations, the wider scientific establishment still has to recognise the substance and importance of this most **'precious'** and sought after of all breakthroughs in science.)

[229] The now extreme soul-corrupted, psychotic state that the human race's immensely heroic 2-million-year long, anger-egocentricity-and-alienation-producing-and-escalating search for knowledge has resulted in was recognised in these, earlier mentioned, lyrics from the band With Life In Mind: **'It scares me to death to think of what I** [we, the whole human race] **have become'**, **'Our innocence is lost'**, **'I** [we] **can't express all the hate that's led me** [us] **here and all the filth that swallows us whole. I** [we] **don't want to be part of all this insanity. Famine and death. Pestilence and war. A world shrouded in darkness'**, **'Everything you've been told has been a lie…We've all been asleep since the beginning of time. Why are we so scared to use our minds?'**, **'Keep pretending; soon enough things will crumble to the ground'**, **'How do we save ourselves from this misery…So desperate for the answers…We're straining on the last bit of hope we have left'**, **'This is the end.'**

[230] And this **'so desperate for the answers'** state that we are in was also recognised in these powerfully honest aforementioned words from R.D. Laing (see paragraph 149): **'We are all murderers and prostitutes…We are bemused and crazed creatures, strangers to our true selves, to one another… the *ordinary* person is a shrivelled, desiccated fragment of what a person can be…our capacity even to see, hear, touch, taste and smell is so shrouded in veils of mystification…Man cut off from his own mind, cut off equally from his own body – a half-crazed creature in a mad world…The outer divorced from any illumination from the inner is in a state of darkness. We are in an age of darkness… between *us* and It** [our true selves or soul] **there is a veil which is more like fifty feet of solid concrete…There is a prophecy in Amos that there will be a time when there will be a famine in the land, "not a famine for bread, nor a thirst for water, but of *hearing* the words of the Lord** [words of truthfulness]**." That time has now come to pass. It is the present age…How do you plug a void plugging a void? How to inject nothing into fuck-all? How to come into a gone world?…I do assure you. The dreadful has already happened'**, and **'We are dead, but think we are alive. We are asleep, but think we are awake. We are dreaming, but take our dreams to be reality. We are the halt, lame, blind, deaf, the sick. But we are doubly unconscious. We are *so* ill that we no longer feel ill, as in many terminal illnesses. We are mad, but have no insight** [into the fact of our madness].**'**

[231] Below is a drawing I have done of the soul-destroyed dystopia apparent on the streets of San Francisco, which is representative of the soul-destroyed dystopia that is happening everywhere in the world! With regard to the unbearable levels of psychosis that I've depicted in the drawing, I've heard it said that the outflow of water to the sea from cities in California is now so full of anti-depressant medication that it's toxic to marine life! Yes, the **'desperate[ly]'** needed neurosis-relieving and psychosis-healing **'answers'** about the human condition ABSOLUTELY HAVE TO BE BROUGHT TO THE ATTENTION OF THE WORLD!

THE DEATH OF HUMANITY

Dystopia of lawlessness, unbearable psychosis, drug addiction and homelessness

[232] In **Berdyaev**'s quote (mentioned in paragraph 193) he said it **'requires great daring'** to find **'knowledge of good and evil'**, which is 'knowledge of the human condition', and this need for **'great daring'** and courage actually applies to two situations. It is needed to confront the subject of the human condition and by so doing find understanding of it, and it is also needed to cope with and support the arrival of the all-redeeming but at the same time all-exposing explanation of the human condition. As I have just said, the corrupting search for the redeeming understanding of the human condition has gone on for so long that there is a great deal of mind upset neurosis and soul-repressed psychosis—basically extreme psychological exhaustion—in the human race. And what this means is that even though the explanation of the human condition is all-redeeming of our corrupted condition, it is also going to be all-exposing of it, and to manage this exposure properly will require **'great daring'** and bravery by the whole now extremely upset human race, most especially by the exceptionally psychologically embattled and insecure sufferers of the human condition.

[233] When **Alvin Toffler** wrote in his famous 1970 book *Future Shock* of **'the shattering stress and disorientation that we induce in individuals by subjecting them to too much change in too short a time'** (p.4 of 505), he was actually anticipating the great shock that the arrival of all-relieving but at the same time all-exposing understanding of the human condition would inevitably initially cause. The following are extreme examples of how difficult it is for some people to cope with the arrival of the exposing truth about the human condition. In a response to our online advertisements (where we have eminent scientists like Professor Harry Prosen promoting our world-saving explanation of the human condition, see www.humancondition.com/online-video-advertisements) one person wrote: **'Your advertisements seriously need to stop. You're hurting people. Every time I hear your annoying advertisements my blood pressure rises. I want you to understand that what you are doing is causing harm to actual people. I've never thought in my wildest dreams that I might be so annoyed by an advertisement. In your publications the endlessly repeated phrase "unconditional selflessness" is evidence of the psychological predatory essence of what you're doing. Your work should be removed and your charitable status be revoked. You're obviously a fraud and I'm not interested, your grandiose claims about solving the human condition are a joke, I cannot stress and repeat that fact enough times, stop this awful nonsense.'** So, even mentioning the reasonably unconfronting and not uncommonly used phrase **'unconditional selflessness'** is too unbearably

confronting for some people. Another truth-hater, and a-future-for-the-human-race-hater, said to me, 'Look, I don't accept that the word "innocent" even exists, and as far as I'm concerned Christ was the most neurotic person who ever lived!'

Last Judgment (detail) by Lieven van den Clite, 1413

[234]The arrival of the all-redeeming but at the same time inevitably all-exposing understanding of the human condition is actually the long-feared 'judgment day' referred to in religion. When the understanding of the human condition is finally found and the 'curtains are drawn open at last', or 'the lights are finally turned on', revealing our 2-million-year corrupted condition, it can't help but come as a great shock. However, because this exposure day, or truth day, or honesty day, or transparency day, or revelation day, is actually a day when compassionate understanding comes to the human race, it is actually a day (a time) of universal relief from excruciating neurological and psychological bondage and suffering. As an anonymous Turkish poet once recognised, judgment day is '**Not the day of judgment but the day of understanding**' (Merle Severy, 'The World of Süleyman the Magnificent', *National Geographic*, Nov. 1987). Humans typically find any significant change difficult to adjust to, and since the arrival of understanding of the human condition brings the biggest change of all

from living in denial of the human condition to living with that denial exposed, that massive paradigm shift was always going to be particularly difficult. There is an inevitable initial great shock of adjustment, especially for those with more psychological upset to have exposed.

[235] What now needs to be explained, and this is obviously very important, is that THERE IS A WAY FOR EVERYONE TO COPE WITH 'EXPOSURE DAY', and that is to not study the explanation of the human condition more than each person's level of soundness of self can cope with. Everyone obviously needs to study the explanation enough to overcome the 'Deaf Effect' and verify it is the explanation of the human condition that we have needed to save the human race, and so deserves everyone's support (in fact everyone's letting-go-of-their-upset-state transformed-way-of-living support), but, beyond undertaking that degree of study of the information, everyone should only study the information as much as their degree of soundness of self can cope with.

[236] This makes sense, just as it made sense for people who weren't exceptionally intellectually clever to accept being in a supportive role in the community to those exceptionally clever people who weren't excluded from university because they were intellectually clever enough to pass the entrance exams and learn how to be successful economists and business managers, etc, and to competently cope with complex subjects like higher physics, abstract mathematics and legal conundrums. As has been explained throughout this book, the unbearably depressing truth of our soul-corrupted human condition, and thus of any acknowledgement of the existence of our species' original cooperative, selfless and loving innocent instinctive self or soul, and thus of any acknowledgement of the world of innocent soundness, has been too unbearably confronting and depressing to admit. So even though innocent soundness is needed to be a successful thinker, the whole notion of soul and its soundness has been ignored, and instead power, fame, fortune and glory achieving intellectual cleverness has been the great focus in human life. All day, every day, the depressing, negative truth about our innocent instinctive soul and of the existence of soundness has been met in the human mind by a firm 'NO, I'm not going there'; while uplifting, positive, intellectual cleverness and smartness has been met by a resounding 'YES, that's what I want to hear'!!

[237] BUT INEVITABLY AND NECESSARILY, with the finding of the human-race-liberating understanding of the human condition this focus changes from being on IQ (intelligence quotient) to being on EQ (emotional

quotient, which is human-condition-avoiding code for soul or soundness quotient, SQ).

[238] Christ predicted this change would happen when understanding of the human condition arrives when he said, **'the meek…inherit the earth'** (Matt. 5:5), and when **'many who are first will be last, and many who are last will be first'** (Matt. 19:30, 20:16; Mark 10:31; Luke 13:30), and when **'The stone the builders rejected has become the capstone'** (Ps. 118:22; Matt. 21:42; Mark 12:10; Luke 20:17; Acts 4:11; 1 Pet. 2:7).

[239] The truth is that ignoring degrees of alienation/soundness in thinking ability, which is what focusing on intellectual IQ and ignoring instinctual EQ has been doing, or, for that matter, focusing on mechanistic not holistic thinking (see paragraph 36), was almost as absurdly dishonest as it would be to have a supposedly meaningful discussion about the ability to see without mentioning the need for eyes, but that degree of absurdity is what has been going on! With such a dishonest focus, no wonder R.D. Laing said (see paragraph 149) that the human race is **'liable to go on to a terminal dementia…The outer divorced from any illumination from the inner is in a state of darkness. We are in an age of darkness.'** (See more about this change of focus from IQ to EQ in paragraphs 657, 1186 and 1188 of *FREEDOM*.)

[240] The description of the basic need and way to support the explanation of the human condition once you have verified that it is the redeeming explanation of the human condition was given in the previous Part 2.9, and is described at length in chapter 9 of *FREEDOM*, and *The Great Transformation*, and also in many other presentations on our World Transformation Movement website, which can all be accessed for free there.

[241] **THE CHOICE FOR THE HUMAN RACE IS VERY CLEAR.**

[242] **EITHER** it denies, ignores, blocks out, attacks, resists this human-race-saving, **'holy grail'** of insight into the human condition and dies in the most horrendously excruciating, absolutely unthinkable state of terminal levels of crippling neurosis and psychosis, which, as I have been emphasising, is already happening everywhere—and can be further emphasised by pointing out that recent generations are being described as 'snowflakes' because they are so psychologically embattled and insecure they melt if placed under any pressure or stress; as a psychiatry journal reported in 2024, **'Young people are showing the most serious warning signs and**

symptoms of a society and a world that is in serious trouble' (McGorry, Patrick D. et al, 'The Lancet Psychiatry Commission on youth mental health', *The Lancet Psychiatry*, Sep. 2024, Vol. 11, Issue 9, 731-774). These further commentaries reveal how deadly serious the situation is: **'when you start thinking about demographics and what's happening with the young people…the Boomer generation has some issues. Gen Xers have a little bit more, millennials have a lot more, Gen Z is going to have way more, and it's just this cascade effect of generations getting worse and worse and worse…when the Boomers are gone, it's going to be apocalyptic…we need their level of stability…Millennials are a psychotic generation…Gen Z is going to be crazier, Alpha is going to be crazier still'** (Tim Pool, *Younger Generations Are Becoming More & More Insane, When They Vote They Will Destroy The US*, YouTube channel 'Timcast IRL', 11 Dec. 2022). And with regard to Generation Alpha: **'Generation Alpha is on track to be the most troubled generation of all time by a long way. This is the era of mindless consumption, extreme polarization, and soulless, meaningless junk food entertainment. From the explosion of mental illness and violent crime to rampant consumerism and alarmism…And Gen A has been placed at the forefront of it all…What if you were born in the last decade? What if you were thrust right into the most unstable and confusing era in human history?…This question encapsulates the one of a kind situation faced by Generation Alpha and no one seems to have an answer to it…The internet is filled with…sensationalist garbage that's designed to hijack your emotions…serve them [children] with a fresh batch of mind-numbing waste… the brain rot of the internet…something that's crafted to be as shocking and stimulating as possible…this works a lot like a drug…Gen A has been exposed to this world from such a young age…As a result, Gen A will likely see a much bigger spike in mental illness than Gen Z, and at a much younger age. Everything from depression to anxiety will be more rampant than ever…and Gen A is the largest generation in human history…a hyper reality will officially come into existence, alongside the beginning of the end of our species…And so far, no one is taking it all that seriously'** (*How Generation Alpha Will Change Society Forever*, YouTube channel 'Novia', 29 Sep. 2023). The core problem with the internet is that it so *massively* facilitates both exposure to the **'shocking'** *horror* of the human condition in the world, and distraction and escape from the personal *agony* of the human condition through the **'soulless, meaningless junk food entertainment'**, **'mindless consumption'**, **'mind-numbing waste'**, alienating **'brain rot'** it provides, that it has *massively* accelerated the psychological upset that has been building ever since humanity's heroic knowledge-finding-but-upsetting search for understanding began some 2 million years ago. The internet has had the corrupting effect of the over-exposure to upset that the development of agriculture had but much, much worse; soul-corruption on steroids! What the upset human mind has so desperately needed exposure to is not **'shocking'** horror upon horror and escapist **'meaningless' 'brain rot'**, but the meaningful 'brain relieving' redeeming, reconciling, healing and **'shock'**-ending *understanding* of the human condition that we now have. Now that we can admit the existence of our original, natural, all-sensitive and all-loving instinctive self or soul, we can see clearly how soul-destroying, dehumanising, alienating modern high-tech communication technology can be. (The fact that I can't type or text, yet have written over 20 books about the deepest of all subjects of the human condition, surely serves as evidence of the benefits of not being hot-wired to modern technology!) And the reason there has been no **'answer to'** this **'apocalyptic'** development and **'no one is taking it all that seriously'** is because the only way we humans have been able to cope with our corrupted condition is by adding more layers of block-out and denial of it — which, again, is a practice that a large part of this book has been dedicated to explaining. As soon as any inference of our corrupted condition comes up we immediately veer away from it. We can't deal with the unbearably depressing truth of our corrupted condition at all. It has been a no-go zone, a red line we won't cross — hence the Deaf Effect that occurs when reading my presentations about the human condition. As R.D. Laing said (see paragraph 149), **'Our alienation goes to the roots'**, our denial of our corrupted condition is *very* highly practised. People are becoming more and more and more psychotic, which actually means soul-dead, but at Resignation we learnt to not allow our minds to go near the suicidally depressing issue of being soul-dead. We can't confront the truth of our corrupted human condition, so we can't look into, **'take' 'seriously'**, the **'apocalyptic'** development that is taking place before our eyes! But no longer is all this crippling, truth-avoiding denial and silence needed because we can at last now face, acknowledge, understand and heal our corrupted condition. So as long as the Transformed Way of Living that the next paragraph emphasises has to be taken up, is taken up, then instead of the internet being **'soulless, meaningless junk food entertainment'**, **'filled with…sensationalist garbage that's designed to hijack your emotions'**, the **'mind-numbing waste'**, **'brain rot of the internet' 'crafted**

to be as shocking and stimulating as possible', which causes **'Everything from depression to anxiety'** to **'be more rampant than ever'**, the internet will be, as I said in paragraph 216, free from all the dishonesty and delusion that made human life so fraudulent, mad, destructive and chaotic. All that dishonest and deluded, immensely destructive existence will be replaced by so much relief, happiness and healthiness that we will find it almost too exciting to bear! Rather than **'Fear…**[being] **driven into our** [their un-resigned, truthful] **minds everywhere we** [they] **look'**, as With Life In Mind sang (par. 169), imagine the difference it will make to children being exposed to all that happiness and excitement on the internet!! (See Part 7 of *Death by Dogma* for more explanation of the origin and description of the terminal levels of psychosis in the world now.)

iStock.com/Todor Tsvetkov

[243]**OR** the human race follows the logic of how everyone can cope with the arrival of the explanation of the human condition by verifying and then supporting the explanation of the human condition while not studying it more than their particular level of soundness of self can cope with. As I have often said, the mantra for the new world that is opening up now that we finally have the redeeming, reconciling and healing understanding of the human condition is, **'If we look after this information it, in turn, will look after each of us and the world.'** That needs to be every-one's orientation now—to take up the Transformed Way of Living and live in support of the explanation while getting on with the main job of

disseminating this human-race-saving understanding and ending all the suffering in the world. This quote attributed to Socrates, that **'The secret of change is to focus all of your energy not on fighting the old but on building the new'**, has great meaning now because we actually have the means, which is the redeeming, reconciling and healing understanding of the human condition, to **'build'** a **'new'**, sane, happy, peaceful world! Basically, the human race must not lose its nerve on the home run, give up at the last moment on fulfilling its great endeavour of finding and then supporting the knowledge required to free itself from the horror of the human condition, namely the biological explanation of the human condition that the human race now has. We have to have the **'great daring'** that Berdyaev counselled and *Don't Stand In The Way, For The Times Are A-Changin'*, as I have titled another book that I have written this year (2025). Yes, since we finally now can, we have to make sure that all the fabulous anticipations that were described in paragraph 213 by Buddha, Moses, Christ, Isaiah, John Lennon and Martin Luther King Jr of the liberation of humanity from the human condition are not outrageously irresponsibly and selfishly stopped from being realised—ensure that the anticipations of Buddha's time of **'Perfect Enlightenment'**, Christ's **'kingdom come'**, Isaiah's ending of the need to **'train for war'**, John Lennon's **'a brotherhood of man'**, and Martin Luther King Jr's time when **'all of God's children, black men and white men, Jews and Gentiles, Protestants and Catholics'** can **'join hands and sing'**, **'Free at last! Free at last! Thank God Almighty, we are free at last!'**, are fulfilled!!

[244] Exposure 'day' or truth 'day' or honesty 'day' *can* be coped with, and in the most fabulously relieving and transforming way where we live in support of the human-race-liberating understanding instead of continuing with the now obsoleted, human-race-destroying practice of living in determined denial of it—which is the fabulous transformation to living a human-condition-free life that was explained in the previous Part 2.9.

[245] Part 2.15 will further describe the fabulous transformation of the human race from living with the horror of the human condition to living free of it. These are four drawings I have made to summarise the fabulously relieving and exciting transformation that every human can now experience that solves the problem of exposure 'day' and avoids the human race dying in excruciating terminal levels of psychosis, levels that are already occurring everywhere!

How to cope with 'Exposure Day' or 'Judgment Day'.
The great transformation of the human race from the horror of the
human condition to living free of it that the redeeming understanding
we now have of the human condition finally makes possible.

[246] Now I have something really, really special for everyone. The following link is to a video of a fellow Australian, the legendary singer John Farnham, performing live in 1989 before a vast audience in Melbourne, Australia, with the Melbourne Symphony Orchestra. He is singing the Beatles' 1965 song *Help!* that was written by, you guessed it, that prophetic genius John Lennon. It is the most real, impassioned and fabulous rendition of the song. And that song is actually an anticipation of the liberation of humanity from the agony of the human condition—it is the **'HELP'** that the young girl pictured on page 143 was so desperately pleading for: **'When I was younger, so much younger than today** [when I was in my innocent youth and when the human race was in its original innocent state] / **I never needed anybody's help in any way / But now these days are gone and I'm not so self assured / Now I find I've changed my mind, I've opened up the doors** [to horrific psychosis within me]. // **Help me if you can, I'm feeling down / And I do appreciate you** [the relieving understanding of the human condition] **being 'round / Help me get my feet back on the ground / Won't you please, please help me? // And now my life has changed in oh so many ways / My independence seems to vanish in the haze / But ev'ry now and then I feel so insecure / I know that I just need you** [that relieving understanding] **like I've never done before. // Help**

me if you can, I'm feeling down / And I do appreciate you being 'round / Help me get my feet back on the ground / Won't you please, please help me? // Help me, help me Ooh.' Link to watch Farnham's performance: www.wtmsources. com/308. As Nick Wilson, the popular online personality who gives his reactions to songs, exclaimed while watching Farnham's performance, 'He is one of the greatest singers I've ever heard!' ('ThatSingerReactions', 31 Aug. 2021). Link to watch Wilson's overwhelmed reaction: www.wtmsources.com/309.

[247] And then there's also *You're The Voice*, Farnham's fabulous un-official Aussie anthem where he sings 'We gotta make ends meet [reconcile the human condition] / Before we get much older' so 'we all can stand together' (1986, Keith Reid et al.). Link to him performing it at the same concert: www. wtmsources.com/311.

[248] I should also mention that Farnham sang with similar immense passion the song *That's Freedom* (Tom Kimmel, 1987), which is also about the need 'to deliver' 'the light in the dark' of understanding of the human condition that 'after the rain' of our suffering will 'rekindle the spark' and 'let freedom ring'—as these lyrics from the song reveal: 'I want something that's open and strong / As the country this road's moving on / From the mountain to the valley / From the ocean to the alley / From the highway to the river / One emotion to deliver / One heart / One way / One love / To share but not to chain / That's freedom / It's a song of the heart / A race in the wind / The light in the dark / That's freedom / It's a reason to live / And after the rain / Rekindle the spark / Let freedom ring / I don't know why it takes so long / To remember what the world's doin' wrong'. Link to watch Farnham singing *That's Freedom* live on his 1990 Australian *Chain Reaction* tour: www.wtmsources.com/310. (Many more prophetic songs about the arrival of understanding of the human condition are included in Freedom Essay 45.)

Part 2.11 "This understanding of the human condition will end all prejudices like racism forever"

[249] What will now be demonstrated is that the Instinct vs Intellect explan-ation demystifies not just the human condition but *all* aspects of human behaviour, however before doing that, what has just been explained in Part 2.10 needs to be emphasised: now that our corrupted human condi-tion has been explained and defended we *can* survive 'exposure day' or 'judgment day' of that corrupted condition. The reality is that the human

race would never have had the tenacity and courage it has had to keep on going with the corrupting search for the redeeming understanding of our corrupted condition if it didn't believe that when that understanding was finally found, as it has now been, we wouldn't be able to cope with the exposure of it—and, as has now been explained, there *is* a way to cope, which is to support the understanding without overly confronting it. The human race *can* now leave Plato's dark cave of denial and live in the warm healing 'sunshine' of redeeming understanding.

[250] Truth is possible now, and that honesty makes a world of relieving difference to *every* human situation. As was described in paragraph 90, the true merits of the right-wing in politics, and the true liabilities of the left-wing, have finally been able to be revealed, and most relievingly the whole stressful business of politics has been brought to an end. As was outlined in paragraph 135, the true heroic role of men and the true naive situation of women have finally been able to be recognised and the historic rift and misunderstanding between men and women is finally able to be reconciled and unified. As outlined in paragraph 61, the religious concept of 'God' has finally been able to be truthfully explained and demystified. And as explained in paragraphs 212-213, being able to explain and end the underlying insecurity in us humans ends the need to have faith in a religion's prophet. Indeed, as mentioned in paragraph 208, being able to heal the corrupted human condition makes possible the real therapy of the human race that all the pseudo idealistic forms of therapy described in *Death by Dogma* have for so long unsuccessfully tried to achieve. In the case of science, as was mentioned in paragraphs 237-239 about EQ/SQ, soundness-recognising holistic, teleological truthful science can replace human-condition-avoiding reductionist, mechanistic, truthless, blind science.

[251] And the human-race-saving redeeming and healing understandings and demystifications go on and on. One more very important demystification needed to save the human race is acknowledgement of the differences in alienation/soul-corruption/loss-of-innocence between 'races' (ethnic groups) of humans. How this truthful acknowledgement ends all prejudices like racism forever is described in chapters 8:16E-F of *FREEDOM*, and is also summarised in Freedom Essay 28, which the following description of how the problem of racism is brought to an end is taken from.

[252] While we couldn't explain the corrupted state of the human condition, one of the most unbearably condemning truths we had no choice but to

live in denial of—even though it is an extremely obvious truth—is that the more and/or longer an individual human, or a race of humans, was exposed to the great battle that humanity has been heroically waging against the ignorance of our instinctive self, the more they would suffer from the psychologically upset angry, egocentric and alienated state of the human condition.

[253] The particular reason this truth of the differences in upset between individuals and races has been intolerable while we couldn't explain, defend and transform our corrupted condition is that it left those more upset feeling, and even portrayed as, less worthwhile and inferior to those who weren't as upset. Now that we can explain and defend the human condition we can see how unjustly condemning and dangerous this prejudice against upset has been. This is because what is revealed by the redeeming understanding of our species' corrupted condition is that while humans do naturally vary in their degree of upset as a result of their different encounters with humanity's heroic but upsetting battle to find that redeeming understanding, ALL HUMANS ARE EQUALLY GOOD. Upset is not a bad, evil state, but a good, heroic one. But again, until we could explain this truth that all humans are equally good there was the danger of unfair and destructive individualist, racist, ageist and sexist prejudice and discrimination against the more upset, and also retaliatory prejudice and discrimination by the more upset towards the less upset for their either direct or implied condemnation of them.

[254] In the relationship between individuals in the social structure of societies we saw in the previous Part 2.10 how the more soul-sound and innocent were ignored and all the emphasis was on intellectual cleverness. Those with high EQ (emotional/soul/soundness quotient) were totally, ruthlessly and murderously-of-soul discriminated and prejudiced against in favour of those with high IQ (intelligence quotient). As Christ was quoted as saying in paragraph 238, the soul-sound, innocent **'meek'** were **'last'** and **'rejected'**.

[255] In the case of the interaction between races, the Holocaust where 6 million Jews were exterminated by the Nazis during World War II, and in more recent times, the attempted 'ethnic cleansing' by the Bantu Hutu of some 800,000 of the tall, elitist Nilotic Tutsi in 100 days of bloodshed in Rwanda in 1994, are two horrific instances of the endless horrific-beyond-description examples in history of the effects of prejudice and discrimination between races.

Nazi propaganda poster proclaiming that 'Jews are lice; they cause typhus'

Cover of Bantu magazine dehumanising Tutsis as 'snakes', 'cockroaches' and 'animals'

[256] It is an immense relief then that understanding of the human condition finally brings an end to all forms of prejudices like racism, because, clearly, without the clarifying explanation for why all humans are equally good, it was *far* too exposing, confronting, condemning, distressing and dangerous to differentiate individuals, races, genders, ages, generations, countries, civilisations and cultures according to how innocent or upset they were. It is no wonder that, to quote an article in *New Scientist*, when ascribing **'genes for negative traits'** to **'ethnic groups'** there has been concerned debate over **'whether some avenues of research are best left un-trodden because what they reveal is bound to be socially and culturally incendiary…Or is it intellectually dishonest, even cowardly, not to investigate all aspects of the human condition?'** (Andy Coghlan, 'Bun fight over warrior gene', *New Scientist* 'Short Sharp Science' blog, 10 Aug. 2006; see www.wtmsources.com/147).

[257] While we couldn't explain and defend the upset state of the human condition it certainly has been **'incendiary'**, *far* too hurtful and dangerous to acknowledge differences in where individuals, races, genders, ages, generations, countries, civilisations or cultures were in their progression from innocence to upset in humanity's heroic journey from ignorance to enlightenment. And so the *extremely* **'dishonest'** attitude of not allowing differentiation simply *had to be* enforced.

[258]HOWEVER, most importantly, now that all this **'dishonest'** denial has been defied and **'all aspects of the human condition'** have been truthfully 'investigate[d]' and, as a result, the explanation of why all humans are equally good has been found and presented in *FREEDOM*, this *situation completely changes*. As emphasised in the previous Parts 2.9 and 2.10, with the defence of upset found, it finally becomes psychologically safe to acknowledge differences in upset—AND ALSO NECESSARY to acknowledge it if we are to truly understand ourselves and stop the march to terminal levels of psychosis and our species' extinction, and to free ourselves from living in Plato's horrible cave of soul-deadening darkness.

[259]The obvious, and in truth only significant behavioural difference that exists between humans is their degree of upset anger, egocentricity and alienation. In the case of races, there are obviously more innocent races who have been relatively isolated from the immensely heroic but immensely upsetting battle humanity has been waging against the ignorance of our instinctive self or soul, and who are thus relatively unadapted to and thus naive about the difficulties of living with the human condition. And there are obviously going to be other races who are more instinctively adapted to upset and can thus manage and cope better with it. And there are obviously going to be other races who are so instinctively adapted to upset that they are too aware of the reality of life under the duress of the human condition and who are thus overly cynical about being ideally cooperatively, selflessly and lovingly behaved, and are thus overly selfish and opportunistic and thus socially uncooperative. The result of all this variation in upset is that some races have been effective in living with the human condition while others have been either too innocent and naive, or too upset, soul-corrupted and cynical.

[260]Again, although we have had to avoid it, it is an obvious truth that humans became increasingly adapted to life under the duress of the human condition, with some races becoming more or less adapted than others. Just as individual humans vary in their degree of alienation from our species' original instinctive, all-loving, all-sensitive, selfless and trusting soulful self, so races of humans naturally vary in their degree of alienation. The longer and/or more intensely an individual or a race of people were subjected to life under the duress of the human condition, the more they naturally became adapted to that upset existence.

[261] While a relatively innocent person or relatively innocent race still behave relatively ideally themselves and expect others to do the same, other individuals and races have become so adapted to the upset/corrupt world that they no longer behave ideally themselves and no longer expect others to behave ideally. The longer humans were exposed to the human-condition-afflicted state the more cynical they became about human existence—a 'cynic' being **'one who doubts or denies the goodness of human motives'**. The psychiatrist Wilhelm Reich wrote honestly about the effects of the different levels of upset in the human race when he described how **'The living** [those relatively free of exposure to upset and thus more soul-alive/innocent]**...is naively kindly...It assumes that the fellow human also follows the laws of the living and is kindly, helpful and giving. As long as there is the emotional plague** [the flood of upset in the world]**, this natural basic attitude, that of the healthy child or the primitive** [innocent race]**...** [is subject to] **the greatest danger...For the plague individual also ascribes to his fellow beings the characteristics of his own thinking and acting. The kindly individual believes that all people are kindly and act accordingly. The plague individual believes that all people lie, swindle, steal and crave power. <u>Clearly, then, the living</u>** [the innocent] **<u>is at a disadvantage and in danger</u>'** (*Listen, Little Man!*, 1948, p.8 of 109).

[262] <u>The consequences for a society of its people becoming overly cynical was that it meant that there would be too little soulful, selfless idealism and too much upset-adapted cynicism-derived selfishness for the society to function effectively</u>. In the situation where it wasn't possible to explain and thus defend upset, the closest people could come to admitting and talking about this fact that people became adapted to the human condition was to describe individuals or families or races or countries or civilisations as having become **'dysfunctional'** and **'decadent'**, and—especially in the case of civilisations—as having **'passed their prime'** or **'peaked'** in terms of their creative powers.

[263] <u>Conversely, some races, like some individual humans, have been too innocent to function effectively in the extremely upset-adapted, human-condition-afflicted, soul-corrupted world</u>. For example, in the extreme, the aforementioned immensely truthful-thinking, prophetic South African philosopher Sir Laurens van der Post described how a member of the relatively innocent Bushmen or San people, some of

whom still survive in the Kalahari Desert of southern Africa, found it impossible to cope with having his innocent, natural spirit compromised: **'You know I once saw a little Bushman imprisoned in one of our gaols because he killed a giant bustard which according to the police, was a crime, since the bird was royal game and protected. He was dying because he couldn't bear being shut up and having his freedom of movement stopped. When asked why he was ill he could only say that he missed seeing the sun set over the Kalahari. Physically the doctor couldn't find anything wrong with him but he died none the less!'** (*The Lost World of the Kalahari*, 1958, p.236 of 253). And Sir Laurens was even more specific when he stated that **'mere contact with twentieth-century life seemed lethal to the Bushman. He was essentially so innocent and natural a person that he had only to come near us for a sort of radioactive fall-out from our unnatural world to produce a fatal leukaemia in his spirit'** (*The Heart of the Hunter*, 1961, p.111 of 233). While still members of the extremely upset stage of humanity, *Homo sapiens sapiens*, the honey-coloured Bushmen are probably the most instinctively/genetically innocent group of people living today. They are more innocent, less soul-corrupted, less adapted to upset, less toughened, than dark-skinned Bantu Africans, but in turn Bantus are not as toughened and thus as operational and successful in the human-condition-afflicted corrupted world as Caucasians from Europe. For example, I once saw a documentary in which a Bantu African said something to the effect that **'My people can't compete with white people, you go to sleep at night only to wake up in the morning to find white people own everything.'** In turn, European Caucasians aren't as cynical, toughened and opportunistic—selfish—as people from even more ancient civilisations who have been involved for the longest amount of time in the upsetting battle humanity has been waging against ignorance, like the Chinese from the ancient Yellow and Yangtze River valley civilisations, the Indians and Pakistanis from the ancient Indus and Ganges River valley civilisations, and the Arabs and Jews from the ancient Tigress, Euphrates and Nile River valley civilisations.

[264] This is a description by leading psychiatrist Dr Clancy McKenzie of how much the less socially and materially successful can envy the more socially and materially successful, which can lead to angry resentment and distress and then breakdown of traditions in those less socially and materially successful societies, and then mass migration by them to the 'successful' countries, which can then lead to social breakdown of those

once 'successful' countries—which are developments that have been greatly accelerated by advances in communication technology: **'While visiting Machu Picchu in Peru in 1979 I noted very poor persons, living in the mountains, who had only the clothes they wore and perhaps a lama or two, but had beautiful, warm smiles and seemed content and happy. Days later I was in Bogota in Colombia. It was a very hot day and we asked the driver to stop at an outdoor tavern to buy cold beer. The people were very impoverished, but there was a TV playing and they were able to view the "outside world" where everyone seemed to have more, and luxury was abundant. I offered to go in with the driver and he urged me to wait in the car. I soon learned why. <u>The absolute hatred was so intense that it was palpable. These people did not have less than those in Machu Picchu but they saw others who had more, and their needs were intensified</u>'** (Letter to Prof. Harry Prosen, 27 Mar. 2006).

[265]Many examples of the tragic consequences of these inevitable differences in adaption to the human condition of different races—of the overly naive, the most operational, and the overly cynical—are given in the aforementioned chapters 8:16E-F of *FREEDOM*, and in the helpful summary of those chapters in Freedom Essay 28, which I strongly recommend reading.

[266]What is so wonderful for the human race now that we have the redeeming, reconciling and human-race-transforming understanding of the human condition, is that EVERYONE CAN LOOK AFTER EACH OTHER NOW ACCORDING TO THEIR ADVANTAGES AND DISADVANTAGES—in a world of truth instead of human-race-destroying lies. No longer will the more innocent and naive have to steal wealth (and leave the rest of their people in poverty) because they are unable to compete with the more upset-adapted; or the more operational who are neither overly innocent or overly upset be resented or envied for their social and material success; or the overly upset who are too cynical and opportunistic have to live either in corruption-endemic, dysfunctional societies or in rigid, freedom-denying, authoritarian, extremely oppressive and brutal societies. Instead, prejudice will go and the human race will live in transformed harmony. As Franklin Mukakanga, a former advertising director and radio host in Zambia, said, **'From my own experience I can say with absolute certainty that, as the word gets out, *FREEDOM* will provide the key to healing poor or strained 'race relations' throughout the world. Basically, <u>this understanding of the human condition will end all prejudices like racism forever</u>'** (2017).

Franklin Mukakanga, founder of the WTM Zambia Centre

[267] Yes, we leave the old effectively dead, dishonest world behind and enter an indescribably happy and free Transformed World, which the songwriter and singer Bono (of the band U2) described when he sang: **'I've conquered my past** [found the dignifying, human race liberating understanding of the human condition] / **The future is here at last** / **I stand at the entrance to a new world I can see** / **The ruins to the right of me** / **Will soon have lost sight of me'** (*Love Rescue Me*, 1988). And which Beethoven's famous *Ninth Symphony* similarly anticipated: **'Joy!'**, **'Joyful, as a hero to victory!'**, **'Join in our jubilation!'**, **'We enter, drunk with fire, into your** [human-condition-understood] **sanctuary…Your magic reunites…All men become brothers…All good, all bad… Be embraced, millions! This kiss** [of understanding] **for the whole world!'** (1824; lyrics from Friedrich Schiller's 1785 poem *Ode to Joy*). And there's also Martin Luther King Jr's stirring speech **'I have a dream'** in which he looked forward to the time when the human race is able to **'allow freedom to ring…from every village and every hamlet, from every state and every city'** because **'all of God's children, black men and white men, Jews and Gentiles, Protestants and Catholics'** can finally **'join hands and sing'**, **'Free at last! Free at last! Thank God Almighty, we are free at last!'** ('I Have A Dream' speech, 28 Aug. 1963).

[268] Also, as that fabulously creative and prophetic Beatle member John Lennon anticipated in his 1971 song *Imagine*—which polls regularly rank

as the best ever written—when he asks us to imagine a world without the condemning differentiation of good and evil, a world liberated from the uncertainty and insecurity of the human condition: **'Imagine there's no heaven / It's easy if you try / No hell below us / Above us only sky** [imagine the end of the duality of good and evil, the reconciliation and amelioration of the human condition] / **Imagine all the people / Living for today / Imagine there's no countries / It isn't hard to do / Nothing to kill or die for / And no religion too** [imagine the world free of the human-condition-produced insecurities that necessitated religious faith, or New Age/PC/Woke deluded dogma, or the egocentric compensations of power, fame, fortune and glory] / **Imagine all the people / Living life in peace / Imagine no possessions / I wonder if you can / No need for greed or hunger / <u>A brotherhood of man</u> / <u>Imagine all the people / Sharing all the world</u> / You may say I'm a dreamer / But I'm not the only one / I hope some day you'll join us / And <u>the world will be as one</u>** [imagine a world free of the human condition and all the resulting alienation].' (Again, more prophetic songs about the arrival of understanding of the human condition are included in Freedom Essay 45.)

'the world will be as one'

Part 2.12 The great danger for humanity of the resistance to the arrival of the explanation of the human condition

[269] All new paradigms in science are typically resisted by the established order. The German philosopher Arthur Schopenhauer summarised the baptism of fire new ideas in science have historically had to undergo when he said that **'the reception of any successful new scientific hypothesis goes through predictable phases before being accepted'**. First, **'it is ridiculed'** and **'violently opposed'** (compiled from two references to Schopenhauer's quote—*New Scientist*, 15 Nov. 1984 and *PlanetHood*, Ferencz & Keyes, 1988). And, as was pointed out in Part 2.10, no new paradigm is going to be **'opposed'** as much as the arrival of the human-race-liberating but at the same time human-condition-exposing explanation of the human condition. It therefore can be expected that truthful analysis and accountable explanation of the human condition will **'begin'** the acceptance of its **'blasphem[y]'** (as Bernard Shaw described the situation initially faced by **'All great truths'**) with extremely hostile and unfair attacks on it and its proponents, **'violent oppos[ition]'** that would have to be survived and overcome if acceptance of the explanation is to develop and the human race be saved from terminal levels of psychosis and extinction. As the science historian Thomas Kuhn warned, **'In science...ideas do not change simply because new facts win out over outmoded ones...Since the facts can't speak for themselves, it is their human advocates who win or lose the day'** (Shirley Strum, *Almost Human*, 1987, p.164 of 297—Strum's references are to Thomas Kuhn's *The Structure of Scientific Revolutions*, 2nd edn, 1970).

[270] And such attacks against this breakthrough, human-race-saving, redeeming, reconciling and healing, fully accountable and thus true, biological Instinct vs Intellect explanation of the human condition and its proponents are precisely what happened. Outrageously false and malicious attacks were made on the explanation and its proponents in the mid-1990s in publications by two of the largest media organisations in Australia, the Australian Broadcasting Corporation and *The Sydney Morning Herald*. Their attacks essentially dismissed this most **'precious in the life of the mind' 'grail'**, as Wilson referred to it, fully accountable, authoritatively supported, Instinct vs Intellect biological explanation of the human condition as meaningless pseudo-science, even making the outrageously false accusation that I am a leader of a sinister organisation. These extremely defamatory attacks had to be, and were, resisted

and defeated. From 1995 to 2010, we in the Sydney WTM Centre—including the eminent, twice-honoured Order of Australia recipient and WTM Patron Tim Macartney-Snape—sought redress and successfully sued both organisations in the Supreme Court of New South Wales for publishing these defamatory lies. You can read about this successful but emotionally and financially exhausting battle, especially for our small charity, in the Persecution of the WTM section on our website, in chapter 6:12 of *FREEDOM*, and in Freedom Essay 56).

Justice David Hodgson (with his Order of Australia insignia)

[271] I would like to mention here that Justice David Hodgson (shown above) wrote the unanimous NSW Court of Appeal decision in the above mentioned successful legal battle that basically saved my Instinct vs Intellect explanation of the human condition from horrifically unjust and outrageously irresponsible persecution by those two of the largest media organisations in Australia. Justice Hodgson was a Rhodes Scholar, a published philosopher and, according to a former Chief Justice of the Federal Court of Australia, **'one of the finest judges who ever graced a court**

in this country' (see www.wtmsources.com/256). Indeed, in his 4 September 2012 obituary in *The Sydney Morning Herald*, Justice Hodgson was described **'from an early age'** as having **'been fascinated by what went on inside the head that gave rise to conscious experience'**, and that he was said to be **'blessed with flawless logic'** and to **'fit the description of Plato's "philosopher king"'** (see www.wtmsources.com/184). <u>Thank you Justice David Hodgson, certainly a true 'philosopher king', from the bottom of my heart.</u>

[272] What Justice David Hodgson did was so precious because, as I said, it saved the denial-free, truthful, human-race-saving, redeeming explanation of the human condition from the destruction that persecution very often leads to. Again, to reinforce Kuhn's warning I included above that **'In science…ideas do not change simply because new facts win out over outmoded ones…Since the facts can't speak for themselves, it is their human advocates who win or lose the day'**, people are sometimes tempted to think that a good idea will withstand whatever resistance it encounters, but that is not true. In John Stuart Mill's 1859 essay, *On Liberty*—a document considered a philosophical pillar of western civilisation—Mill emphasised this point when he wrote that **'the dictum that truth always triumphs over persecution is one of those pleasant falsehoods which men repeat after one another till they pass into commonplaces, but which all experience refutes. History teems with instances of truth put down by persecution. If not suppressed for ever, it may be thrown back for centuries'** (*American state papers; On liberty; Representative government; Utilitarianism*, 1952, p.280 of 476).

[273] There is another human-race-saving example of denial-free honesty and truthfulness being saved by a member of the legal profession. It was by a lawyer named Gamaliel during the birth of Christianity. His counsel was essentially for people to trust in the principles of democracy to ascertain what is of value to humanity and what is not, rather than to resort to dishonest slander, vicious abuse and persecution. The apostles had been gaoled and were threatened with death for defending Christ's extraordinarily truth-defending, denial-free teachings and life. (As I mentioned earlier in paragraph 152, you can read my description of Christ's human-race-saving, truth-defending life in Freedom Essay 39—about which many people have, for example, said, **'This is the best description of Christ I have ever read!'**.) To quote from the Bible about what happened to the apostles: **'a Pharisee named Gamaliel, a teacher of the law, who was honoured by all the people, stood up in the Sanhedrin [the full assembly of the elders of Israel] and ordered that the men [apostles] be put outside for a little**

while. Then he addressed them: "Men of Israel, consider carefully what you intend to do to these men. Some time ago Theudas appeared, claiming to be somebody, and about four hundred men rallied to him. He was killed, all his followers were dispersed, and it all came to nothing. After him, Judas the Galilean appeared in the days of the census and led a band of people in revolt. He too was killed, and all his followers were scattered. Therefore, in the present case I advise you: Leave these men alone! Let them go! For if their purpose or activity is of human origin** [i.e., is a product of deluded, insecure egocentricity and alienation]**, it will fail. But if it is from God** [if it is sound]**, you will not be able to stop these men; you will only find yourselves fighting against God." His speech persuaded them. They…Let them go'** (Acts 5:34-38). This was human-race-saving counsel because it allowed the Christian movement to survive its most vulnerable early stage when, after Christ's death, only a few people were aware of the immense importance of the movement and where the death of those few could very well have meant the end of the movement—and, as I explain in my essay about the importance of Christ's contribution to the world, the death of humanity.

[274] I again need to emphasise that our World Transformation Movement is not a religion. We are supporting the actual biological understanding that makes the ideal life possible, whereas religions were only able to support the embodiment of idealism in the form of the sound prophet the religion was founded around. However, while we are not a religion, what we are presenting is like a religion in that it is based on supporting denial-free honesty/truth/substance/soundness/authenticity.

[275] To return to what happened to us. It took 15 long years of legal battles for the outrageously dishonest 1990s attacks to be defeated, and, as a result, for the explanation of the human condition that the human race has been tirelessly working towards and dreaming of finding to be allowed to live, and for support of it to grow.

[276] And that support *has* grown since those dreadful years when we had to endure horrific persecution. At the time of writing, which is mid-2025, the WTM's private Facebook Group has over 80,000 members and there are 80 WTM Centres around the world. Also, support from sound-thinking, esteemed scientists and thinkers keeps growing. You can see many commendations from such scientists and thinkers at www.humancondition.com/reviews-commendations, but to just include two here that have already been mentioned. In 2016 one of the world's leading psychiatrists, Professor Harry Prosen, a former President of the Canadian

Psychiatric Association, said, '**I have no doubt that Jeremy Griffith's biological, instinct vs intellect explanation of the human condition is the holy grail of insight we humans have sought for the psychological rehabilitation of our species.**' And in 2024, the eminent ecologist from San Diego State University, Professor Stuart Hurlbert, said: '**I am stunned and honored to have lived to see the coming of "Darwin II". I say this because after Darwin's theory of Natural Selection explained the variety of life, Jeremy Griffith has gone on to solve the other four main questions science had to answer about our world and place in it. They are: 1) the dilemma of the human condition, which his instinct vs intellect explanation in chapter 3 of his main, seminal book *FREEDOM* finally solves; 2) how we humans became fully conscious when other species haven't, which he answers in chapter 7; 3) the origins of humans' unique moral nature, which he answers in chapter 5, which it turns out American philosopher John Fiske had already explained in 1874 but mechanistic science had ignored; and 4) the truth of the Integrative Meaning of existence (which we have personified as 'God'), in chapter 4, which only a rare few thinkers in history have been able to recognize. And having been able to solve those primary issues he has, in chapter 8, using first principle and fully accountable biological explanations been able to resolve all the secondary problems like: the polarized state of politics; the rift between men and women; the schism between science and religion; the conflict between individuals and between races (thus ending aggression and war at its source); and, above all, bring an end to the threat of terminal psychosis and our species' extinction! A truly phenomenal, beyond description, scientific achievement!**'

[277] Unfortunately, our defeat of persecution hasn't lasted because in the mid 2020s, Artificial Intelligence (AI) technology, which is fast gaining a reputation for spreading all manner of defamatory untruths—including so-called '**AI hallucinations**'—is dredging up and even fabricating more of these old discredited lies, requiring us pioneering supporters of this human-race-saving understanding of the human condition to again have to redress these totally false and utterly irresponsible attacks—including, if necessary, again suing the perpetrators for defamation.

[278] I have seen AI accused of '**churning out bile and falsehoods with…alarming fluency…of creating a firehose of conspiracy, grievance and hate online. If you train an AI on that, it will repeat what it learns. Garbage in, garbage out…AIs…display total ignorance of how historical truth is established…AIs…are not independent thinkers. They are elaborate parrots…We are not training**

Large Language Model AIs simply to distil wisdom, we are letting them binge on our noise' (Jonathan Sacerdoti, 'Grok's fascist screed was no one-off', *The Spectator*, 10 Jul. 2025). Yes, AI's **'churning out [of] bile and falsehoods'** in **'total ignorance of how historical truth is established'** creates a very serious problem when the breakthrough biological explanation of the human condition arrives. AI won't have the necessary capability, knowledge or wisdom to make proper sense of what is being presented or to see through the blatant prejudice of its opponents, and will instead rely excessively on all that **'bile and falsehoods'** in making its so-called 'assessments'!

[279] As mentioned, the other publication I have written this year (2025) to accompany this *The Human Condition* publication is *Don't Stand In The Way, For The Times Are A-Changin'*, and the reason it was written was specifically to warn people against standing in the way of the liberation of humanity from the human-race-destroying horror of the human condition by attacking this fully accountable and authoritatively supported biological explanation of the human condition with outrageously false lies—and also to try to warn online platforms like Meta and Google that they are committing this worst possible cybercrime where AI propagates these enormously dangerous and utterly irresponsible lies. This is the illustration with its text that I did for the cover of that book to emphasise the extreme irresponsibility of standing in the way of the liberation of humanity from the horror of the human condition.

Don't 'stand in the way' by being a truth-hating murderer of a future for humanity

Drawing by J. Griffith © 2014-2025 Fedmex Pty Ltd

Part 2.13 Another instance of Goya's ability to reveal the truth about the human condition

[280] What follows in the remaining sections of Part 2 are further examples of how fearful of the truth of our horrifically soul-corrupted condition the human race has been, and as a result, how determinedly we have presented a human-condition-denying representation of ourselves. And also how honest and exposing of all that dishonesty a very rare few people in history have been—like Sir Laurens van der Post, Plato, Nikolai Berdyaev, R.D. Laing, Michael Leunig, Francis Bacon, Edvard Munch, Francisco Goya, Vincent van Gogh, William Blake, Bob Dylan, and the other presenters of honest representations of the human condition that are included in this book.

[281] To refer to Goya again (see paragraph 187), in addition to his etching *The sleep of reason brings forth monsters*, he also painted two contrasting representations of a famous festival in Spain, one denying the human condition and the other revealing it. The esteemed Australian *TIME* magazine art critic, Robert Hughes, described these pictures in his 2002 documentary *Goya: Crazy Like A Genius*: '**There are two paintings of the same subject…**[They are of] **a big religious festival, that of St. Isidro. On that day thousands of citizens, in their Sunday best, converged on a pilgrimage chapel outside Madrid and had a picnic.'** In the first representation titled '*St. Isidro's Meadow*' [below], Hughes said, '**the girls are in their white parasols, the men in their finery, the scene is of social pleasure and jollity'**.

Goya's *St. Isidro's Meadow* (detail), 1788

[282]Then, according to Hughes, **'Thirty years later Goya returned to the same theme. In this picture [below, titled]…*The Pilgrimage of St. Isidro*, instead of these happy fashionable well-dressed young people, you have this horrible snake of…dark figures…like demons crawling across an ash heap. The faces are…of madmen and hysterics…The whole picture is deeply threatening'**.

Goya's *The Pilgrimage of St. Isidro* (detail), 1821-1823

[283]While Hughes wasn't sound and secure enough to go beyond saying **'The faces are…of madmen and hysterics'** and recognise that Goya was depicting the horrifically soul-corrupted state of our human condition, Goya clearly knew humanity was living a completely fraudulent, human-condition-denying, escapist, deluded existence. Accompanying his etching, *Capricho 6*, he even wrote that **'The world is a masquerade. Looks, dress and voice, everything is only pretension. Everyone wants to appear to be what he is not. Everyone is deceiving, and no one ever knows himself.'** So Goya was an extremely bold, human-condition-confronting, prophetic, truthful thinker.

Part 2.14 The incredible, beyond-description bravery of the human race

[284]Since we humans are now so practiced at denying the existence of our corrupted condition, and putting on a brave face and presenting a buoyant representation of ourselves to the world, as is being done by the people in Goya's first picture, it is very difficult for us to know how much we suffer from the corrupted state of the human condition. Without the contrasting pictures like Goya's, and the other revealing portrayals of our corrupted condition that have been included, we wouldn't be able to see just how absolutely incredibly brave and heroic we humans have been. But now that we can finally compassionately understand our species' whole tragic and horrific story we can begin to appreciate just how astronomically courageous we humans have really been coping with the human condition for some 2 million years. The Biblical prophet Isaiah described our horrific situation truthfully when he said: '**justice** [redeeming understanding] **is far from us, and righteousness does not reach us. We look for light, but all is darkness; for brightness, but we walk in deep shadows. Like the blind we grope along the wall, feeling our way like men without eyes...Truth** [i.e. understanding of our corrupted condition] **is nowhere to be found'** (Isa. 59). And as a prophet of our time, and a Nobel Laureate for Literature, Bob Dylan, sang about our desperately estranged existence, '**How does it feel to be on your own, with no direction home, like a complete unknown'** (*Like a Rolling Stone*, 1965).

[285]Yes, now that we can finally understand our immensely courageous human journey we can begin to appreciate how horrible it was when we started to corrupt the innocent, loving existence we once lived in but had no idea at all why we were committing such a seemingly unforgivable crime. The shame has been absolutely immense, and the loneliness and wretchedness of our existence absolutely terrible! The great English artist William Turner's painting *Fishermen at Sea* captures something of the phenomenal heroism of the human race for struggling for 2 *million years* through this terrible, terrible lonely darkness of guilt-stricken bewilderment and seeming evil badness, and the feeling that left us with that we are no-good, utterly meaningless, worthless creatures!

J.M.W. Turner's *Fishermen at Sea*, 1796

[286]It is a very powerful metaphorical depiction of our condition. Surrounded by a terrifying dark, surging storm of condemnation, people are hunkered down in a small boat trying to look after each other against the overwhelming horror of their situation. We had to be our own friends because we were no longer a friend of our soul and the rest of the natural world associated with it. <u>We humans have in truth been very, very, *very* alone beings, but no longer because we can now return home to the happy cooperative, selfless and loving state we once lived in—that is as long as we don't stand in the way of the fabulous transformation of the human race from living with the horrific suffering of the human condition to living free of the human condition and all those horrific effects!</u>

Part 2.15 "We gotta help the people out, right now", and that is what we can all do now!

[287] All these exceptionally honest thinkers and artists that have been mentioned in this book have been *extremely* helpful in evidencing all that is being explained—for which myself and all of us in the WTM are eternally grateful. However, after all the confronting truth that we have been wading through in this presentation that explains the human condition, it would be good to return to discussing how fabulously exciting the world for humans can be now that the fabulous transformation of the whole human race from living with the agony and horror of the human condition to living free of it can occur.

[288] Given the serious plight of the human race now, it is firstly highly appropriate to include these next four paragraphs from what I said last year in my 2024 presentation *Sermon On The Beach* about the timeliness of the human-race-redeeming and transforming new way of living for the human race that understanding of the human condition finally makes possible.

[289] The world *is* becoming so horribly messed up and dysfunctional that now, despite our historic practice of maintaining a positive, everything-is-okay outlook, more and more people are being able to let that delusion go and accept that this world-disintegrating situation can't go on and that a fundamental change *has to occur*. As Professor Harry Prosen wrote in his Introduction to my book *FREEDOM*, '**I think the fastest growing realization everywhere is that humanity can't go on the way it is going.**' Just recently, the renowned podcaster Joe Rogan was talking about the world having reached this completely exhausted state where we desperately need a new, more loving and honest ideology that a human-condition-confronting-not-avoiding, truthful thinker like Jesus Christ would present, saying, '**I think as time rolls on people are going to understand the need to have some sort of belief in the sanctity of love and truth. And a lot of that** [those values] **comes from religion…Ethics are based on our moral compass and we all have one, but that's not necessarily true** [anymore], **we need Jesus!**' (7 Feb. 2024). This led to another well-known podcaster, Russell Brand, agreeing with Rogan, saying, '**Yes, I do think we need a return of Jesus Christ…We need clarity, honesty, open-mindedness and a real alternative in the political sphere… I'd certainly like to see some new independent movements challenging these old institutions**' (7 Mar. 2024).

[290]Two years earlier, in 2022, in one of his podcasts, the Scottish historian and archaeologist Neil Oliver expressed identical sentiments about how the political Right and Left have become meaningless for people and they are looking elsewhere, saying, **'So it's not Left and Right, it's become right and wrong, good and bad. For a lot of people…they're trying to understand what the rules** [for a new ideology] **might be, and where help might be found. So people are invoking myths like** [the prophesised return of King] **Arthur and people who had faith in** [the prophesised 'second coming' of] **Jesus Christ, both those stories take you to the same place that, "cometh the hour, cometh the man". People are looking to and hoping for a time when an Arthur or a Jesus will come back to right the wrong, it's amazing! Many people all around the world are feeling betrayed and let down by those** [leaders and institutions] **to whom previously they would've looked for safety. It's suddenly like finding a loving parent revealed as the opposite of a loving parent, an abusive parent, and people's faith in the structures and the institutions of society have been rocked and for a lot of people that faith might never come back, and they're looking to put their faith in something else. I think that's why you get a lot of people talking about alternatives, because a lot of people want to start again, a lot of people want to start alternative institutions'** (3 Sep. 2022). In 2023 the YouTuber and author Kevin A. MacLean similarly pleaded for the return of King Arthur at the end of his documentary *Who Are The Welsh?*, saying, **'The need for Arthur's return couldn't be greater, and I'll be watching for him'** (28 Jan. 2023). I talk about how the finding of the human-race-saving redeeming explanation of our corrupted human condition represents the fulfilment of the prophecy of the return of King Arthur-like, relatively innocent, Celtic, soul-strong, denial-defiant courage and leadership in paragraph 1036 of *FREEDOM*, and about the fulfilment of the prophecy of the 'second coming' of Christ-like, innocent, sound and loving truthfulness in paragraphs 1278-1279 of *FREEDOM*, and in Freedom Essay 39.

[291] So, despite all the historic denial, more and more people are waking up from all the delusion and denial that has owned human life and are admitting this end play, human-condition-stricken situation *has* now arrived. And once a person understands this information, understands that the human condition has been explained, they actually know that there is another way we can live; they actually know that **'a real alternative' 'new independent movement'** based on **'the sanctity of love and truth'** that Rogan and Brand called for, and an **'alternative institution', 'not Left and Right'** but based on **'right and wrong, good and bad'** values that Oliver called

for, *is* available, which is *exactly* what the Transformed Way of Living represents.

²⁹²Yes, the truth absolutely is that the human race has reached this end play situation where the old aggressive, dishonest, human-condition-denying, power and glory way of living is finished with, it's no longer viable, and we need an alternative, and, blow me down, there *is* an alternative that is all-solving and all-wonderful, so everyone *has* to let go of that old absolutely exhausted and absolutely obsoleted way of living and take up the alternative Transformed Way of Living—and when they clearly face and accept that truth and as a result actually decide to, and do, take up that way of living, they will, at that moment, suddenly experience how *incredibly* relieving and exciting it is to be living in the all-solving and all-exciting Transformed Way of Living! In paragraphs 1198-1200 of *FREEDOM* I described how when St Paul let go his attachment to his resigned dishonest way of living and decided to live in support of Christ, his 'conversion experience' of relief was so great he fell off his donkey and went blind for three days. Well, as I mentioned earlier in paragraph 217, now that we can live in support of the actual human-race-transforming understanding of the human condition, the relief and happiness is so incredibly great it is like falling off the back of a thousand donkeys and of our eyes being so flooded with tears of happiness we won't be able to see out of our eyes for months!

Drawing by J. Griffith © 2023 Fedmex Pty Ltd

St Paul falling off his donkey and going blind with ecstatic relief
after he let go of his struggle with the human condition

[293]Earlier in paragraph 188 the rock band U2's 1997 song *Staring At The Sun* described how unbearably confronting the metaphorical light of the sun has been of our corrupted condition, but the sun is also an appropriate metaphor for the liberating light of understanding of our corrupted condition, which is why it has been used in many songs that anticipate the arrival of that redeeming understanding. In Freedom Essay 45 I present a collection of these immensely inspiring songs which, for example, includes Jimi Hendrix's 1970 song *Hey Baby (New Rising Sun)* from his wonderfully titled album *First Rays Of The New Rising Sun* (cover below). In the song, Hendrix imagines encountering a radiant figure that is **'gonna spin and spread around peace of mind; and a whole lotta love to you and you!'**. He asks her **'where do you come from?'**, and she says **'the land of the new rising sun** [a human-condition-free world]'. Hendrix then gives voice to the urgent need to resolve the human condition: **'we're gonna go cross the Jupiter's sands, and see all your people one by one! We gotta help the people out, right now; that's what I'm doing here, all about.'** The song concludes with Hendrix's plaintive appeal: **'Please take me** [to this fabulous place of freedom from the human condition]**!'**

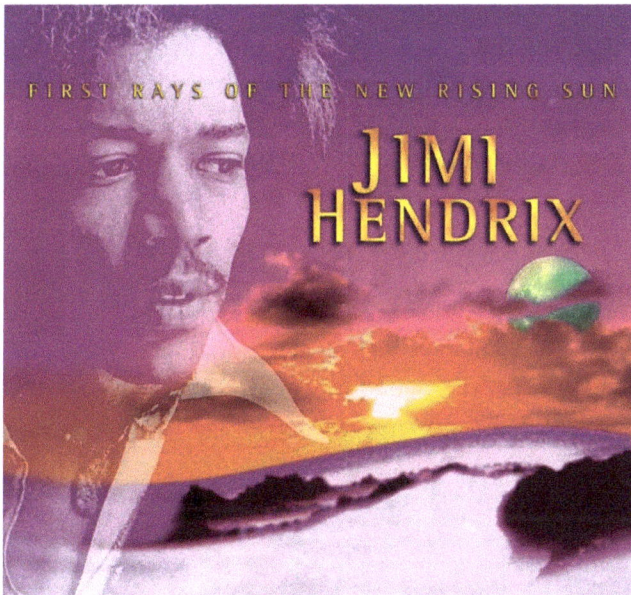

Sony/Legacy Recordings

[294]Well, it is absolutely true that **'We gotta help the people out, right now'**, AND THAT'S WHAT WE CAN ALL NOW DO. I will include here again the

words of WTM founding member Tony Gowing and the poet T.S. Eliot from the end of Part 1, and now add the counsel of R.D. Laing, the honest words of Christ, and the insight of author and philologist J.R.R. Tolkien.

[295] Tony Gowing said about the fabulous, dreamed-of-and-now-coming-true, great transformation of the human race that **'We don't have to be ashamed. We don't have to shake our fists at the heavens anymore and prove to everyone and everything that they have been wrong about us. First-principle science has proven that we are worthwhile; that we are gloriously heroic beings. It had to be the way it's been—there was no other way—but it's all over now. The relief of finally being able to understand floods through our whole being; the anger and frustration dissipates; all the bullshit, falseness and lies end. We can finally love ourselves and participate in the world instead of constantly fighting it. No longer preoccupied with proving our self-worth, we will finally have the room in ourselves to properly help others; to selflessly participate in stopping the suffering everywhere we look.'**

[296] Yes, this is the dream of the awesome transformation that R.D. Laing was referring to (see paragraph 149) when he wrote, **'True sanity entails in one way or another the dissolution of the normal ego, that false self competently adjusted to our alienated social reality: the emergence of the "inner"** [unresigned, denial-free, soulful] **archetypal mediators of divine power, and through this death a** [transformed] **rebirth, and the eventual re-establishment of a new kind of ego-functioning, the ego now being the servant of the divine, no longer its** [necessary] **betrayer.'** And **'an intensive discipline of un-learning is necessary of *anyone* before one can begin to experience the world afresh, with innocence, truth and love'**.

[297] To describe the fabulous transformation that is now available to everyone using Christ's honest description (see paragraph 152) of our horrific human-condition-stricken state: this is the end of having **'baptized your souls in the water of darkness'**, and being **'drunk with the** [angry] **fire and full of bitterness'**, and of having **'walked by your own** [deranged] **whims'**.

[298] This is the end of our being **'soaked with the sense of exile'** from **'Eden'** that J.R.R. Tolkien recognised that we **'long for'** when he wrote, **'We all long for Eden, and we are constantly glimpsing it: our whole nature at its best and least corrupted, its gentlest and most human, is still soaked with the sense of exile'** (*The Letters of J.R.R. Tolkien*, ed. Humphrey Carpenter, 1981).

[299] And yes, finally, at last, T.S. Eliot's anticipation of our return home has become a reality—back to a state of human-condition-free sanity,

peace, togetherness and happiness, but this time as fully conscious beings. As he described our heroic journey, **'We shall not cease from exploration and the end of all our exploring will be to arrive where we started and know the place for the first time.'**

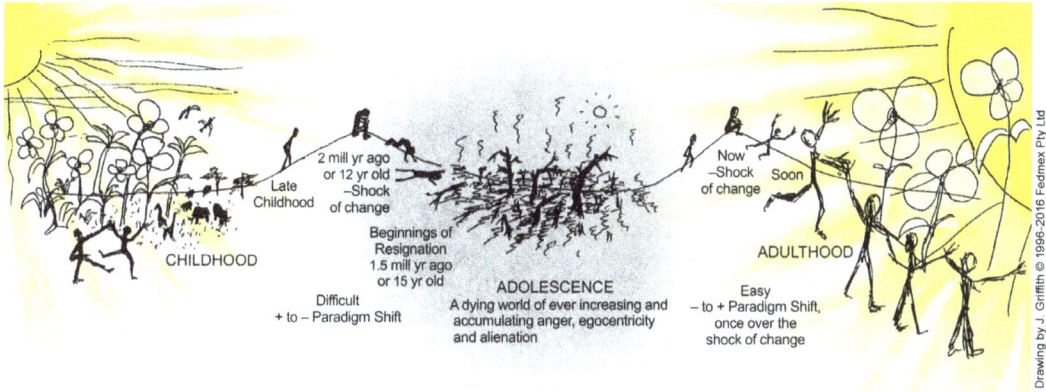

Humanity's Journey from Ignorance to Enlightenment, see chapter 8 of *FREEDOM*.

[300] So that ends this description of what the human condition really is, and of the transformed world that understanding of it makes possible. In my book *The Great Guilt that causes the Deaf Effect*, and in Video/Freedom Essays 1, 2, 3 and 4, and especially in Video/Freedom Essays 10 and 11, you can read/watch more descriptions and explanations of what the human condition is.

Part 2.16 "Since new ideas in science can't speak for themselves, it is their human advocates who win or lose the day"

[301] Since, as the science historian Thomas Kuhn warned (see paragraph 269), **'In science…ideas do not change simply because new facts win out over outmoded ones…Since the facts can't speak for themselves, it is their human advocates who win or lose the day'**, the support that this project of bringing scientific understanding to the historically most forbidden of all subjects—but **'most precious in the life of the mind' 'grail'**, as Wilson referred to it—of the human condition, has been world-saving in its importance.

[302] Therefore, in addition to the support and endorsements from numerous distinguished thought leaders, the ever-growing base of support that our work at the World Transformation Movement (WTM) is receiving from our already, in mid-2025, 80,000 Facebook Group members, and especially from our 80 WTM Centres that have already been established globally, and most especially from our 50 Founding Members of the WTM who have fought so valiantly all these years to defend our human-race-saving, Instinct vs Intellect explanation of the human condition, is so precious they are actually more precious than human language is capable of describing!

[303] The following are the names of these especially heroic and precious 50 Founding Members, who, incidentally, once they knew of the importance of the information and of their support, even made the extraordinary decision not to have children in order to devote every possible resource they have to their mission of saving the human race. (To be clear, this practice of not having children is not something we advocate for all members of the WTM, it was simply a personal initiative we founding members felt we had to take once we understood the difficulty and seriousness of our undertaking where we had to make sure we gave ourselves every possible chance of succeeding in pioneering support for the human-race-saving explanation of the human condition): Annabel Armstrong, Susan Armstrong, Sam Belfield, John Biggs, Richard Biggs, Anthony Clarke, Lyn Collins, Steve Collins, Lachlan Colquhoun, Sarah Colquhoun, Eric Crooke, Emma Cullen-Ward, Fiona Cullen-Ward, Anthony Cummins, Neil Duns, Sally Edgar, Anna Fitzgerald, Brony FitzGerald, Connor FitzGerald, Tony Gowing, Jeremy Griffith, Simon Griffith, Damon Isherwood, Felicity Jackson, Charlotte James, Lee Jones, Monica Kodet, Anthony Landahl, Doug Lobban, Tim Macartney-Snape, Manus McFadyen, Tony Miall, Rachel O'Brien, James Press, Professor Harry Prosen, Stacy Rodger, Marcus Rowell, Genevieve Salter, Nick Shaw, Wendy Skelton, Ali Watson, Polly Watson, Prue Watson, Tess Watson, Tim Watson, Annabelle West, James West, Stirling West, Prue Westbrook and Annie Williams.

[304] This is a photograph that was taken on 17 November 2024 in Sydney of 'the rising sun' formed by many of our Founding Members, including other precious WTM Members, some from other WTM Centres, and also some friends of the WTM!

L to R: Front row: Doug Lobban, Ales Flisar*, Virginia Topete*, Marcus Rowell, Polly Watson, Brony FitzGerald; Second row: Annie Williams, Molly van Hemert*, Jeremy Griffith, Lachlan Dunn*, Nicoletta Akritidis*, Nikola Tsivoglou*, Andreas Mavridis*, Chris Akritidis*, Genevieve Salter, Susan Armstrong, Tess Watson, James Press; Standing third and fourth rows: John Biggs, Wendy Skelton, Anthony Landahl, Monica Kodet, Nick Shaw, Stacy Rodger, Anthony Cummins, Dr Carol Velásquez*, Emmα Cullen-Ward, Ali Watson, Fiona Cullen-Ward, Tony Gowing, Sarah Colquhoun, Neil Duns, Stirling West, Prue Westbrook, Lachlan Colquhoun, Connor FitzGerald, Prue Watson, Felicity Jackson, Damon Isherwood, Tony Miall, Tim Macartney-Snape, James West, Sam Belfield, Tom Shannon*, Anna Fitzgerald, Eric Crooke. *Other WTM Members or friends of the WTM

Part 3
Addendum from the WTM

[305] For an appreciation of Jeremy Griffith's redeeming explanation of the human condition you can read the many commendations from thought leaders and from the general public at www.humancondition.com/reviews-commendations. And we think everyone would particularly enjoy this powerful endorsement of Jeremy's presentations about the human condition from Canadian WTM supporter Michael Perry. You can listen to Michael's short audio titled *The Truth Is* at www.humancondition.com/the-truth-is, and/or read the following transcript of what Michael says.

Michael Perry, Alberta, Canada, May 2025

Transcript

[306] Michael's Introduction to his video: "I'd like to share some words I wrote while thinking about the new world that is now possible for us humans. I see more and more that this [WTM] Facebook group is much more than a Facebook group. It's an online resistance to denial. World-Savers, Planet-Healers. It's absolutely clear that biologist Jeremy Griffith has unlocked humanity's will to save the world."

[307] Video: "I've recently been inspired to share a little bit of something that I wrote regarding the new fabulous, magnificent world that's on its

way because of this information. So I'd like to share this, apologies if you hear some large trucks in the background driving by my house!

[308] This is called *The Truth Is*.

[309] The truth is: This knowledge explains what we are. It does so with undeniable, verifiable and irresistible proof. Proof which is rooted and threaded with the soundest, most logically digestible knowledge at the deepest, most fundamental level imaginable. A level that is in truth measureless and limitless, boundless and fearless. We're not even currently capable or mentally stable enough to truly grasp just how different the world is gonna be one day because of this knowledge.

[310] The truth is: Humanity's mind will now open up and it will live every day in its new paradigm of *knowing* it's worth. Every single human will be completely saturated with meaning. Divine, astronomically beautiful, heroic beings are walking this Earth and those beings are us humans. We've just been asleep that's all, waiting for this moment to come. A caterpillar transforms into a butterfly, a cold dark night transforms into a sunrise, a species in denial transforms into a species that lives in the truth of its condition, then transforms into a species that is free of that condition. A species which can finally see itself for the *drop dead astonishment* that it is. That we all are.

[311] The truth is: Soon we will know with all of our might and heart how meaningful and special this world is. And most especially how meaningful it is to be a conscious animal. And how important love is in the survival of that animal. We will know what we are and what it all means like we know water is wet. The heart of humanity is finally jumping in the driver's seat because it has discovered its true potential and it's all because biologist Jeremy Griffith has synthesized a miracle, biological understanding of *what we are*.

[312] The truth is: This knowledge calmly and reassuringly tells the dying world of denial "try to prove me wrong, *I fucking dare you*". It is this knowledge which will bring the dying world, along with us humans in this current state BACK TO LIFE. It is here to stay and *it has won the day*.

[313] The truth is: This knowledge defends what we are with an impenetrable, denial resistant shield of honour and bravery made from the highest calibre known to this world, which is *truth*. This all new, exciting, relieving, loving shield of truth is here for the rest of this Earth's life. The shield's fibres and structures are made with real compassion and empathy from the purest source around—which is humanity's love.

Our love—our original instinctive orientation to this world—the most astonishingly beautiful thing to have ever been developed in our known universe, that is, up until finding THIS understanding, which is now THE *most astonishingly* beautiful thing to have ever been developed and found in our known universe. We used to be the only animal that lied. <u>Now we become the most glorious flood of truth this world has *ever known*</u>.

[314] Thank you humanity for never giving up. Thank you for this meaning. <u>We all have tremendous, tremendous meaning now. Each person who understands and experiences the effects of this information now has the best job anyone could ever dream of—which is distributing this to the rest of our species</u>.

[315] Thank you biologist Jeremy Griffith for digging up the truth at last and assembling it in such a way, the only damn way that could ever make sense. And perfect sense at that.

[316] The truth is: Jeremy Griffith is the real and urgently needed, glaringly obvious, completely legitimate—Prince of Peace.''

[317] Absolutely yes Michael, as Hendrix wrote and sang, **'We gotta help the people out, right now'**, and as you say, **'the truth is'** we actually now CAN, so everyone come and join us in transforming the human race, and let's 'fix the world' because at last we REALLY can!

With the real problem of the human condition finally solved we can now ACTUALLY fix the world!

[318] <u>And the following discussions in the WTM's Facebook Group wonderfully evidence Michael saying</u>, **'this knowledge which will bring the dying world, along with us humans in this current state BACK TO LIFE…**

This all new, exciting, relieving, loving shield of truth is here for the rest of this Earth's life…Now we become the most glorious flood of truth this world has ever known…We all have tremendous, tremendous meaning now. Each person who understands and experiences the effects of this information now has the best job anyone could ever dream of—which is distributing this to the rest of our species.'

[319] By way of background, in July 2025 the WTM decided to have an annual Fix The World Day on the first Saturday of October. While all the Facebook Group posts about the inaugural Fix The World Day gatherings that took place in many parts of the world on 4 October 2025 are amazing to read, the following post by Karen Boon from the WTM Staffordshire (UK) Centre most perfectly evidences that Jeremy's vision that he has stated in every one of his books going right back to his first book *Free* in 1988—which is that **'soon from one end of the horizon to the other will appear an army in its millions to do battle with human suffering and its weapon will be understanding'**—IS COMING TRUE!

[320] We want to emphasise that throughout history there have been an endless stream of movements that people have excitedly joined hoping the movement would bring about a more cooperative and loving soulful existence—Jeremy's book *Death by Dogma* documents the sequence of the main movements that have occurred. However, as is pointed out in that book, these were all FALSE STARTS to ending the agony and horror of the corrupted state of the human condition, because *only* understanding of that corrupted condition could end it. As the Swiss subscriber to the WTM who runs Shamanic Healing workshops said (see paragraph 153): **'My healing work isn't possible without Jeremy's biological explanation of the human condition because no matter how much you connect people with the soul, the mind still questions and needs to understand *why* we carry so much guilt, and Jeremy's work is the medicine for the mind to make the connection between mind and soul.'** What you will notice is happening in these Fix The World gatherings, which is clearly apparent in the following post, is that the REAL START to a soul-resuscitating, sound, genuinely loving and happy life is occurring. You will see how the explanation of the human condition ACTUALLY **'is the medicine for the mind'** that satisfies humans' **'questions and needs to understand'** and by so doing ACTUALLY brings people together in a REAL way!

[321] <u>So this is Karen Boon's 5 October 2025 fabulous, signs-of-a-whole-new-world post</u> (with some extra photos added):

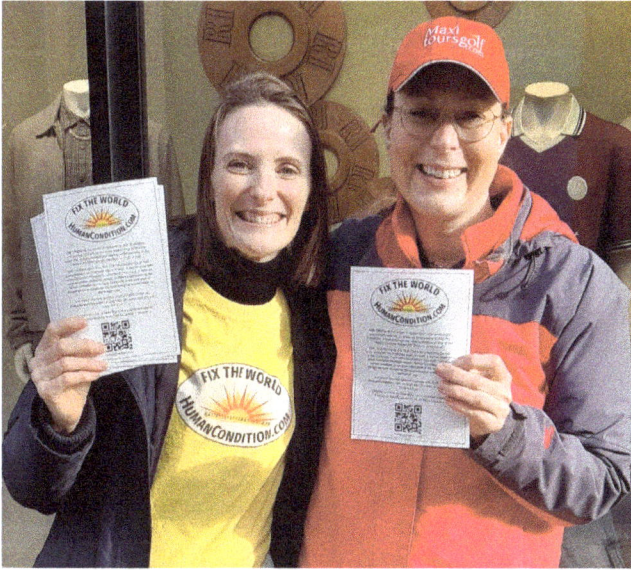

Karen Boon (left) and Maria Ingham at Staffordshire's (UK) Fix The World Day

[322] "Maria Ingham and I spent the first 'FIX THE WORLD DAY' yesterday in Stafford, England handing out leaflets, and it was absolutely amazing!! We had an unbelievably good time spreading the good news, a lot of people accepted leaflets (some said no but that was ok), and a few people here and there stopped to ask what it was all about, which we gladly then discussed with them. Maria & I felt so comfortable in each other's company—even though we had literally just met and Maria hadn't previously handed any leaflets out, etc before—she took to it like a duck to water, she was so impressive! The conversation between us instantly flowed which also helped with our leaflet distribution, because we felt supported and encouraged by each other and our understanding of Jeremy's work. It was such a pleasure to spend time distributing this world-saving information to people, we were both so astonished that we had the power to fix the world in our hands, it was truly amazing and we both thoroughly enjoyed it. What a rewarding way to spend your time! We grabbed a hot drink afterwards and had

some more heartfelt, meaningful conversation before we parted ways feeling all-inspired and ready to carry on the good work going forwards. Thanks so much Kevin Ryan [WTM Dublin] for your wonderful inspiration and initiative to hand out leaflets in the centre of Dublin [which in July 2025 led to Claire Rickie (WTM Kent) inviting everyone to join Kevin and her and make it an event; and then Ales Flisar (WTM Sunshine Coast) suggested it become an annual event; which Ari Akritidis (WTM Melbourne) then suggested we call 'The Fix The World Day'; which led to the creation of our global 'Fix The World Day']—we both said how impactful it is to be face to face with people giving this information out—I even hugged a couple of people I'd spoken to! Anyway, it's been really heartwarming to see so many different initiatives and get togethers around the world for this first Fix The World Day we've just had, I can only imagine how much this event is going to grow every year and how exciting this will all become. Wooo hooo, let's go! 💛☀️💛"

Kevin Ryan at Dublin's Fix The World Day

[323] KAREN'S POST THEN PRODUCED A CASCADE OF INSPIRED COMMENTS:

[324] Stefan Rössler: "This is so exciting to see you together Karen and Maria, and I'm really getting emotional reading your closing words about

how **'it's been really heartwarming to see so many different initiatives and get togethers around the world for this first fix the world day we've just had, I can only imagine how much this event is going to grow every year and how exciting this will all become.'** YES!! This is so exciting and it's only going to grow and grow and grow and grow and grow, just like Kevin said to us in Dublin yesterday! ☀️😻"

Dublin's Fix The World Day (L to R: Claire Rickie, David Brown, Kevin Ryan, Angela Ryan, Stefan Rössler, Yvonne Hayes, Jack Soden)

[325] "Karen Boon: "Stefan this was so lovely! 😊It made me think of Tony Gowing saying in Video/Freedom Essay 5 that: **'I'm telling you, really and truly, that after the initial shock of exposure, the overwhelming effect that this great transition will be known for is *excitement*. When you cotton on to this, I guarantee you will be jumping out of your skin with excitement, just can't-stop-dancing excitement. Honestly, this is all so freeing of your life—in fact, *so* relieving of any and every situation—the excitement is nearly too much to bear! This fabulous breakthrough will transport you and everyone else to a completely wonderful new existence. We aren't bad, we are wonderful beings, we are glorious. There *is* a happy ending, there *is* a get-out-of-jail-free card here—love is all around, boom crash bam wham, the human condition is over—Let's Go!!'** Yay! 🥳🎉🌅"

Tess Watson and Tony Gowing at Sydney's Fix The World Day

326 **Maria Ingham**: "Stefan ah, it was so heartwarming to see you and all the wonderful WTM family gathering together and setting those sparks of truth alight for the wonderful people of Dublin. We got emotional too, it's just the most incredible thing that's happening. 🤗💛"

327 **Desi Akritidis**: "Spot on Stefan, each year momentum will just grow and grow grow!!!"

Melbourne's Fix The World Day, including Ari and Desi Akritidis
(second from left, and in the centre in pink)

[328] Maria Ingham: "It was the best day, truly! I just felt so comfortable with Karen Boon straight away, as if we'd known each other for ages, but we literally had only just met on FTW day. It was the biggest joy of joys to meet another person from the WTM in the flesh and to hug and laugh and talk about deep and meaningful things together while spreading the joy of what Jeremy has given to the world. It was just the most fabulous experience and even though we both had very cold hands, we just couldn't tear ourselves away from carrying on giving out leaflets, knowing what impact that very next leaflet could potentially have on helping another person. 🥹🤩 We've definitely got the bug and are already planning our next meet up to hand out more flyers!! 💕 Just a thoroughly wonderful day in wonderful, smiley, heartfelt company 💛🙏🥰"

[329] Kevin Ryan: "Isn't it just Absolutely Amazing, Maria, the connection right away, it's absolutely out this World Amazing 🐬🐬🐬"

[330] Kimberly Prickett: "Maria that is so heart-warming to read. To have this common interest to be part of the WTM and also to feel an instant connection because you both have a mutual purpose—that's powerful 🥰"

[331] Maria Ingham: "It's incredibly special Kimberly. 🥹💛"

[332] John Nichol: "What a fantastic FTW day post Karen Boon 🎉. It's only going to get bigger and better! Fix the 🌍"

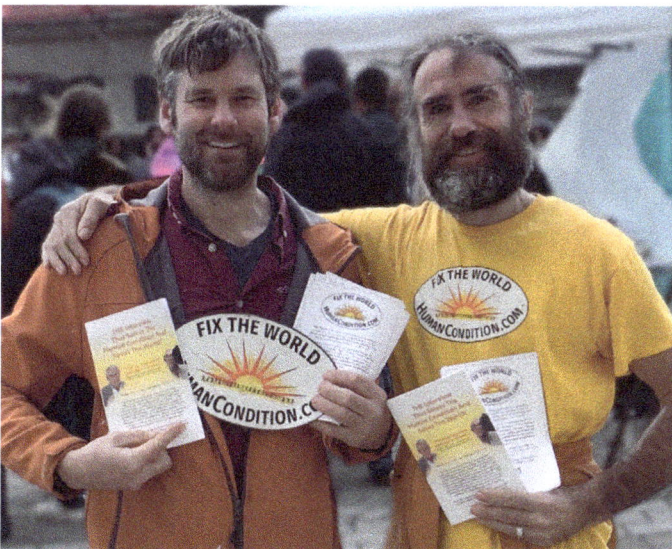

Nick Rule and John Nichol at Cornwall's (UK) Fix The World Day

[333] Kevin Ryan: "Beyond Beyond Beyond Beyond Beyond Beautiful, it's the WTM Family 🌞🌞🌞🌞🌍"

[334] Karen Boon: "Kevin yay, where we are all together united in understanding of our worth, at last 😬💛🦋"

[335] Maria Ingham: "Karen **'we are all united in understanding of our worth, at last'**, oh it's huge isn't it? This is, and will, smash to smithereens all boundaries, all seeming differences and bring together everyone from anywhere and everywhere. That is so bloody special, you can't buy that sort of preciousness. 🤩"

[336] Karen Boon: "Maria oh it's amazing isn't it! I can't quite believe it—I always think of Jeremy saying in a WTM advert: **'It doesn't matter who you are, having rational, biological understanding of the human condition will completely transform your life into the most wonderful existence imaginable. This is because being able to understand the human condition gives you the ability, for the first time ever, to understand every aspect of human behaviour in yourself and in everyone else with 100% clarity and confidence.'** And it's SO amazing that that is all it takes for humanity to come together in peace—real (biological) understanding about ourselves which is what this information provides! 🙏🌟 "

[337] Maria Ingham: "Karen great quote from Jeremy that you added there. It seems so simple yet it's absolutely true! 💕 "

[338] Jules Jamieson: "Best Day ever 🤎 "

[339] Claire Rickie: "What a wonderful day you & Maria had! It sounds perfect—sharing this understanding, putting it on its stand & meaningful conversation 💜👌 "

[340] Maria Ingham: "Claire **'perfect'** is absolutely the best word to describe it. I can't wait for the next FTW Day already!! We're not going anywhere, we will only get stronger and louder until the sparks catch fire and the truth spreads wildly round the whole world 🤎🤎🤎"

[341] Claire Rickie: "Maria absolutely Maria! People just won't be able to resist joining in on the fun! 💛 "

[342] Yvonne Hayes: "I love seeing that photo of you both looking so happy and relaxed in each other's company, whilst sharing the most wonderful news ever to be seen on planet earth!"

[343] Cindy McCaugherty: "It really is so special isn't it Karen and Maria when we get together with others who also have this understanding! You begin to get a glimpse, a sense of what the New World that is coming is going to be like!"

Cindy and Bill McCaugherty's stand at Canada West Coast's Fix The World Day

[344] Maria Ingham: "Cindy oh yes, 💗 we can scarcely imagine the enormity of the goodness that is coming to everyone, but even if it feels like the connection we had on FTW day, every day, then that would be unbelievably out of this world. 😊🤩 "

[345] Monica Kodet: "Pure joy jumping out of the screen 💛💛 so special 🤩 "

Sydney's Fix The World Day, with Monica Kodet (third from left)

[346] **Ales Flisar**: "I was thinking the same thing Karen, how much it will pick up every year, how the truth about us humans and the effects of it will transform each individual. I can imagine how nice the get together must have been, conversation on a completely different level now that we have the crux solved."

[347] **Maria Ingham**: "Ales yes absolutely. Imagine meeting someone for the first time and launching into deep and meaningful conversations right from day 1 😃. No more swerving around the gigantic elephant in the room, but straight to the actual things that really matter in life!! It's a dream come true for someone who has struggled with small talk all my life 😃😊🥰xx "

[348] **Karen Boon**: "Ales & Maria, it's the absolute best thing ever! Real, deep and meaningful conversation with this understanding is at last possible, which means our deeper instinctive selves start to heal and reap the benefits of all the honesty and realness and love that's possible. It's so special, thank goodness a million times the human condition is over 💛💛💛"

[349] AND THE COMMENTS CONTINUED IN THAT VEIN.

[350] YES, *LET'S GO* TO A WHOLE NEW WORLD FREE FOREVER OF THE AGONY AND HORROR OF THE HUMAN CONDITION!!